T0220550

Practical Python AI Projects

Mathematical Models of Optimization Problems with Google OR-Tools

Serge Kruk

Apress®

Practical Python AI Projects: Mathematical Models of Optimization Problems with Google OR-Tools

Serge Kruk
Rochester, Michigan, USA

ISBN-13 (pbk): 978-1-4842-3422-8   ISBN-13 (electronic): 978-1-4842-3423-5
https://doi.org/10.1007/978-1-4842-3423-5

Library of Congress Control Number: 2018934677

Managing Director, Apress Media LLC: Welmoed Spahr
Acquisitions Editor: Steve Anglin
Development Editor: Matthew Moodie
Coordinating Editor: Mark Powers

Cover designed by eStudioCalamar

Cover image by Freepik (www.freepik.com)

Distributed to the book trade worldwide by Springer Science+Business Media New York, 233 Spring Street, 6th Floor, New York, NY 10013. Phone 1-800-SPRINGER, fax (201) 348-4505, e-mail orders-ny@springer-sbm.com, or visit www.springeronline.com. Apress Media, LLC is a California LLC and the sole member (owner) is Springer Science + Business Media Finance Inc (SSBM Finance Inc). SSBM Finance Inc is a **Delaware** corporation.

For information on translations, please e-mail rights@apress.com, or visit www.apress.com/rights-permissions.

Apress titles may be purchased in bulk for academic, corporate, or promotional use. eBook versions and licenses are also available for most titles. For more information, reference our Print and eBook Bulk Sales web page at www.apress.com/bulk-sales.

Any source code or other supplementary material referenced by the author in this book is available to readers on GitHub via the book's product page, located at www.apress.com/9781484234228. For more detailed information, please visit www.apress.com/source-code.

Printed on acid-free paper

A Chloé et Laurent.

Table of Contents

About the Author

Serge Kruk, PhD is a professor at the Department of Mathematics and Statistics at Oakland University and worked for Bell-Northern Research. His current research interests still bear the stamp of practicality enforced by years in industry: algorithms for semidefinite optimization, scheduling, feasibility, and the related numerical linear algebra and analysis. After a few wandering years studying physics, computer science, engineering, and philosophy in Montreal in the seventies, the author entered the industrial world and spent more than a decade designing optimization software, telecommunication protocols, and real-time controllers. He left Bell-Northern Research, the best geek playground in Canada, to become the oldest student in the faculty of Mathematics of the University of Waterloo and attach the three letters "PhD" to his name. The intention, at first, was to return to the real world. But a few years misspent as mathematics and computer science instructor at Waterloo, Wilfrid-Laurier, and finally Oakland convinced him of the appeal of academia. Since then he has wandered as far geographically as Melbourne and as far culturally as Ile de la Réunion, mostly teaching and consulting, with the occasional foray into research and guiding a couple of doctoral students through the painful process of dissertation.

About the Technical Reviewer

 **Michael Thomas** has worked in software development for more than 20 years as an individual contributor, team lead, program manager, and vice president of engineering. Michael has more than 10 years of experience working with mobile devices. His current focus is in the medical sector, using mobile devices to accelerate information transfer between patients and health care providers.

Acknowledgments

This book, and my writing process in general, whether of programs or prose, would not be possible without free software, software for which I have the source, software I can read and modify, software I can trust. First and foremost, Emacs, TeX, LaTeX, and org-mode, but also Python, or-tools, Graphviz, Linux, FreeBSD, and the whole GNU zoo. Thank you to all who participate in the free software movement.

CHAPTER 1

Introduction

1.1 What Is This Book About?

Artificial intelligence is a wide field covering diverse techniques, objectives, and measures of success. One branch is concerned with finding provably optimal solutions to some well-defined problems.

This book is an introduction to the art and science of implementing *mathematical models* of *optimization problems*.

An optimization problem is almost any problem that is, or can be, formulated as a question starting with "What is the best ... ?" For instance,

- What is the best route to get from home to work?

- What is the best way to produce cars to maximize profit?

- What is the best way to carry groceries home: paper or plastic?

- Which is the best school for my kid?

- Which is the best fuel to use in rocket boosters?

- What is the best placement of transistors on a chip?

- What is the best NBA schedule?

© Serge Kruk 2018
S. Kruk, *Practical Python AI Projects*, https://doi.org/10.1007/978-1-4842-3423-5_1

These questions are rather vague and can be interpreted in a multitude of ways. Consider the first: by "best" do we mean fastest, shortest, most pleasant to ride, least bumpy, or least fuel-guzzling? Besides, the question is incomplete. Are we walking, riding, driving, or snowboarding? Are we alone or accompanied by a screaming toddler?

To help us formulate solutions to optimization problems, optimizers[1] have established a frame into which we mould the questions; it's called a model. The most crucial aspect of a model is that it has an objective and it has constraints. Roughly, the objective is what we want and the constraints are the obstacles in our way. If we can reformulate the question to clearly identify both the objective and the constraints, we are closer to a model.

Let's consider in more detail the "best route" problem but with an eye to clarify objective and constraints. We could formulate it as

> Given a map of the city, my home address, and the address of the daycare of my two-year-old son, what is the best route to take on my bike to bring him to daycare as fast as possible?

The goal is to find among all the solutions that satisfy the requirements (that is, paths following either streets or bike lanes, also known as the constraints) one path that minimizes the time it takes to get there (the objective).

Objectives are always quantities we want to maximize or minimize (time, distance, money, surface area, etc.), although you will see examples where we want to maximize something and minimize something else; this is easily accommodated. Sometimes there are no objectives. We say

---

[1] I use the term "optimizers" to name the mathematicians, theoreticians, and practitioners, who, since the nineteen-fifties, have worked in the fields of linear programming (LP) and integer programming (IP). There are others who could make valid claims to the moniker, chiefly among them researchers in constraint programming, but my focus will be mostly in LP and IP models, hence my restricted definition.

that the problem is one of feasibility (i.e. we are looking for any solution satisfying the requirements). From the point of view of the modeler, the difference is minimal. Especially since, in most practical cases, a feasibility model is usually a first step. After noticing a solution, one usually wants to optimize something and the model is modified to include an objective function.

1.2 Features of the Text

As this text is an introduction, I do not expect the reader to be already well versed in the art of modeling. I will start at the beginning, assuming only that the reader understands the definition of a variable (both in the mathematical sense and in the programming sense), an equation, an inequality, and a function. I will also assume that the reader knows some programming language, preferably Python, although knowing any other imperative language is enough to be able to read the Python code displayed in the text.

Note that **the code in this book is an essential component**. To get the full value, the reader must, slowly and attentively, read the code. This book is not a text of recipes described from a birds-eye view, using mathematical notation, with all the nitty-gritty details "left as an exercise for the reader." This is implemented, functional, tested, optimization code that the reader can use and moreover is encouraged to modify to fully understand. The mathematics in the book has been reviewed by mathematicians, like any mathematical paper. But the code has been subjected to a much more stringent set of reviewers with names Intel, AMD, Motorola, and IBM.[2]

---

[2]My doctoral advisor used to say "There are error-free mathematical papers." But we only have found an existence proof of that theorem. I will not claim that the code is error-free, but I am certain that it has fewer errors than any mathematical paper I ever wrote.

The book is the fruit of decades of consulting and of years teaching both an introductory modeling class (MOR242 Intro to Operation Research Models) and a graduate class (APM568 Mathematical Modeling in Industry) at Oakland University. I start at the undergraduate level and proceed up to the graduate level in terms of modeling itself, without delving much into the attendant theory.

- Every model is expressed in Python using Google OR-Tools[3] and can be executed as stated. In fact, the code presented in the book is automatically extracted, executed, and the output inserted into the text without manual intervention; even the graphs are produced automatically (thanks to Emacs[4] and org-mode[5]).

- My intention is to help the reader become a proficient modeler, not a theoretician. Therefore, little of the fascinating mathematical theory related to optimization is covered. It is nevertheless used profitably to create simple yet efficient models.

- The associated web site provides all the code presented in the book along with a random generator for many of the problems and variations. The author uses this as a personalized homework generator. It can also be used as a self-guided learning tool.

  `https://github.com/sgkruk/Apress-AI`

---

[3]`https://github.com/google/or-tools`
[4]The one and only editor: `http://emacs.org`
[5]`http://orgmode.org/`

1.2.1 Running the Models

There is danger in describing in too much detail installations instructions because software tends to change more often than this text will change. For instance, when I started with Google's OR-Tools, it was hosted on the Google Code repository; now it is on GitHub. Nevertheless, here are a few pointers. All the code presented here has been tested with

- Python 3 (currently 3.7), although the models will work on Python 2

- OR-Tools 6.6

The page `https://developers.google.com/optimization` offers installation instructions for most operating systems. The fastest and most painless way is

```
pip install --upgrade ortools
```

Once OR-Tools are installed, the software of this text can be downloaded most easily by cloning the GitHub repo at

```
git clone https://github.com/sgkruk/Apress-AI.git
```

where the reader will find a Makefile testing almost all the models detailed in the text. The reader only has to issue a make to test that the installation was completed successfully.

The code of each section of the book is separated into two parts: a model proper, shown in the text, and a main driver to illustrate how to call the model with some data. For instance, the chapter corresponding to the set cover has a file named `set_cover.py` with the model and a file named `test_set_cover.py` which will create a random instance, run the model on it, and display the result. Armed with these examples, the reader should be able to modify to suit his needs. It is important to understand that the mainline is in `test_set_cover.py` and that file needs to be executed.

1.2.2 A Note on Notation

Throughout the book, I will describe algebraic models. These models can be represented in a number of ways. I will use two. I will sketch each model using common mathematical notation typeset with T$_E$X in math mode. I will then express the complete, detailed model in executable Python code. The reader should have no problem seeing the equivalence between the formulations. Table 1-1 illustrates some of the equivalencies.

Table 1-1. *Equivalence of Expression in Math and Python Modes*

Object	Math Mode	Python Mode
Scalar Variable	X	X
Vector	v_i	v[i]
Matrix	M_{ij}	M[i][j]
Inequality	$x + y \leq 10$	x+y <= 10
Summation	$\sum_{i=0}^{9} x_i$	sum(x[i] for i in range(10))
Set Definition	$\{i^2 \mid i \in [0, 1, \ldots, 9]\}$	[i**2 for i in range(10)]

1.3 Getting Our Feet Wet: Amphibian Coexistence

The simplest problems are similar to those first encountered in high school: the dreaded word problems. They are algebraic in nature; that is, they can be formulated and sometimes solved using the simple tools of elementary linear algebra. Let's consider here one such problem to illustrate the approach to modeling and define some fundamental concepts.

A zoo biologist will place three species of amphibians (a toad, a salamander, and a caecilian) in an aquarium where they will feed on three different small preys: worms, crickets, and flies. Each day 1,500 worms, 3,000 crickets, and 5,500 flies will be placed in the aquarium. Each amphibian consumes a certain number of preys per day. Table 1-2 summarizes the relevant data.

Table 1-2. *Number of Preys Consumed by Each Species of Amphibian*

Food	Toad	Salamander	Caecilian	Available
Worms	2	1	1	1500
Crickets	1	3	2	3000
Flies	1	2	3	5000

The biologist wants to know how many amphibians, up to 1,000 of each species, can coexist in the aquarium assuming that food is the only relevant constraint.

How to we model this problem? All optimization and feasibility problems in this book are modeled using a three-step approach. We will expand on this approach as we encounter problems on increasing complexity, but the fundamental three steps remain the cornerstone of a good model.

1. **Identify the question to answer.** This identification should take the form of a precise sentence involving either counting or valuating one or more objects. In this case, how many amphibians each species can coexist in the aquarium? Notice that "How many amphibians?" would not be precise enough because we are not interested in the total count, but rather in the count of each species. Formulating a precise question is often the hardest part.

Once we have this precise question, we assign a variable to each of the objects to count. We will use x_0, x_1, and x_2. These are traditionally known as *decision variables*. The expression is a misnomer in our first example but reflects the origins of optimization problems in logistics where the decision variables were indeed representative of quantities under the control of the modeler and mapped to planning decisions.

2. **Identify all requirements and translate them into constraints.** The constraints, as you will see throughout the book, can take on a multitude of forms. In this simple problem, they are algebraic, linear inequalities. It is often best to write down each requirement in a precise sentence before translating it into a constraint. For the coexistence case, the requirements, in words, are

   - All amphibians combined consume 1,500 worms.

   - All amphibians combined consume 3,000 crickets.

   - All amphibians combined consume 5,000 flies.

   Note that a statement starting with "The amount of ..." may not be precise enough. In our simple case, there are no specified units but there could be. For instance, the amount consumed could be stated in grams while the availability is in kilograms. This happens often and is the cause of many a model going awry.

Yet, even with our seemingly precise statements, there is an ambiguity left to consider. It is one of the main contributions of a good modeler to highlight ambiguity and clarify problem statements. Here, do we mean that the amphibians will consume exactly the amounts stated, or that they will consume at most the amounts stated?[6] We will assume that "at most" is the proper form of the requirement, both because it is more interesting and, in a sense, subsumes the "equal" question. We will then translate these requirements into algebraic constraints based on our decision variables.

Let's consider worms. The toads eat two per day. The salamanders and caecilians each eat one. Since we decided on x_0 toads, x_1 salamanders, and x_2 caecilians, the total number of worms consumed will be bounded by the following inequality:

$$2x_0 + x_1 + x_2 \leq 1500 \tag{1.1}$$

Had we decided that "equal to" was the proper constraint, we would replace the inequality by an equality.

---

[6]This seemingly trivial change from "exactly equal" to "at most" represents more than 2,000 years of mathematical development in solution techniques. We have known how to solve the "equal" form since ancient Babylonians (though it is known today as "Gaussian elimination") and we teach it in high school, but we only discovered how to solve the "at most" form in the twentieth century.

Consider now crickets. Toads consume one per day while salamanders consume three and caecilians consume two. They will collectively consume $x_0 + 3x_1 + 2x_2$ and we obtain the constraint

$$x_0 + 3x_1 + 2x_2 \leq 3000 \tag{1.2}$$

The constraint on flies is obtained similarly to produce

$$x_0 + 2x_1 + 3x_2 \leq 5000 \tag{1.3}$$

3. **Identify the objective to optimize.** The objective is, in the case of an optimization problem, what we want to maximize (or minimize). In the case of a feasibility problem, there is no objective, but in practice, most feasibility problems are really optimization problems that have been incompletely formulated.

Since the problem is stated as "How many amphibians of each species can coexist?", a possible, even likely, reading is that we want the maximum number of amphibians. (The minimum number is zero and is an example of the uninteresting *trivial* solution.) In terms of our decision variables, we want to maximize the sum and obtain

$$\max x_0 + x_1 + x_2 \tag{1.4}$$

At this point we have a model! Not *the* model, but *a* model: a simple, clear, and precise algebraic model that has a solution, one that answers our original question.

Since we are not mere theoreticians uninterested in practical applications, our next step is to solve the model. As we will do for every model in this book, we need to translate the mathematical expressions above ((1.1)-(1.4)) into a form digestible by one of the many solvers available.

Over the years, optimizers have developed a number of specialized modeling languages and solvers. Here is a short list of the better-known ones:

- Modeling languages
  - AMPL (www.ampl.com)
  - GAMS (www.gams.com)
  - GMPL (http://en.wikibooks.org/wiki/GLPK/ GMPL (MathProg))
  - Minizinc (www.minizinc.org/)
  - OPL (www-01.ibm.com/software/info/ilog/)
  - ZIMPL (http://zimpl.zib.de/)
- Solvers
  - CBC (www.coin-or.org/)
  - CLP (www.coin-or.org/Clp/)
  - CPLEX (www-01.ibm.com/software/info/ilog/)
  - ECLiPSe (http://eclipseclp.org/)
  - Gecode (www.gecode.org/)

- GLOP (`https://developers.google.com/optimization/lp/glop`)

- GLPK (`www.gnu.org/software/glpk/`)

- Gurobi (`www.gurobi.com/`)

- SCIP (`http://scip.zib.de/`)

We should maintain a distinction between *modeling languages*, formal constructions with specific vocabulary and grammars, and *solvers*, software packages that can read in models expressed in certain languages and write out the solutions, although in some cases this distinction is blurry.

As a modeler, one creates a model (in language X) which is then fed to a solver (solver Y). This can happen because solver Y knows how to parse language X or because there is a translator between language X and another language, say Z, which the solver understands. This, over the years, has been the cause of much irritation ("What? You mean that I have to rewrite my model to use your solver?").

To make matters worse, these languages and solvers are not equivalent. Each has its strengths and weaknesses, its areas of specialization. After years of writing models in all the languages above and then some, my preference today is to eschew specialized languages and to use a general-purpose programming language, for instance Python, along with a library interfacing with multiple solvers. Throughout this book I will use Google's Operations Research Tools (OR-Tools), a very well-structured and easy-to-use library.

The OR-Tools library is comprehensive. It offers the best interface I have ever used to access multiple linear and integer solvers (MPSolver). It also has special-purpose code for network flow problems as well as a very effective constraint programming library. In this text, I will display only a very small fraction of this cornucopia of optimization tools.

One of the many advantages of using a general purpose language like Python is that we can do the modeling part as well as the insertion of the models into a larger application, maybe a web or a phone app. We can also easily present the solutions in a clear format. We have all the power of a complete language at our disposal. True, the specialized modeling languages sometimes allow more concise model expression. But, in my experience, they all, at one point or another, hit a wall, forcing the modeler to write kludgy glue to connect a model to the rest of the application. Moreover, writing OR-Tools models in Python can be such a joy.[7] The whole coexistence model is shown at Listing 1-1.

Listing 1-1. Amphibian Coexistence Model

```
1    from ortools.linear_solver import pywraplp
2    def solve_coexistence():
3        t = 'Amphibian coexistence'
4        s = pywraplp.Solver(t,pywraplp.Solver.GLOP_LINEAR_
             PROGRAMMING)
5        x = [s.NumVar(0, 1000,'x[%i]' % i) for i in range(3)]
6        pop = s.NumVar(0,3000,'pop')
7        s.Add(2*x[0] + x[1] + x[2] <= 1500)
8        s.Add(x[0] + 3*x[1] + 2*x[2] <= 3000)
9        s.Add(x[0] + 2*x[1] + 3*x[2] <= 4000)
10       s.Add(pop == x[0] + x[1] + x[2])
11       s.Maximize(pop)
12       s.Solve()
13       return pop.SolutionValue(),[e.SolutionValue() for e in x]
```

---

[7]Writing in Common Lisp would be even better. Alas, there is no Lisp binding for OR-Tools yet.

Let's deconstruct the code. Line 1 loads the Python wrapper of the linear programming subset of OR-Tools. Every model we write will start this way. Line 4 names and creates a linear programming solver (hereafter named s) using Google's own[8] GLOP. The OR-Tools library has interfaces to a number of solvers. Switching to a different solver, say GNU's[9] GLPK or Coin-or[10] CLP is a simple matter or modifying this line.

On line 5, we create a one-dimensional array x of three decision variables that can take on values between 0 and 1000. The lower bound is a physical constraint since we cannot have a negative number of amphibians. The upper bound is part of the problem statement as the biologist will not put more than 1,000 of each species in the test tube. It is possible to state ranges as any contiguous subsets of $(-\infty, +\infty)$, but, as a general rule of thumb, restricting the range as much as possible during variable declaration tends to help solvers run efficiently. The third parameter of the call to NumVar is used as the name to print if and when this variable is displayed, for instance, in debugging a model. We will have little use for this feature as we prefer to write bug-free models.

The constraints on lines 7 to 9 are direct translations of the mathematical expressions (1.1)-(1.3). The order of the terms is irrelevant. In contrast to some restrictive modeling languages, we could have written line 7 as

```
1500>=x[0]+x[2]+x[1]
```

or

```
x[0]+x[1]+x[2]-1500<=0
```

or any other equivalent algebraic expression.

---

[8]https://developers.google.com/optimization/lp/glop
[9]www.gnu.org/software/glpk/
[10]https://projects.coin-or.org/Clp

At line, 6, we declare an auxiliary variable, pop. Though there is no such distinction in the modeling language, this is not a decision variable but rather a helpful device to model the problem. We use this auxiliary on line 10 where we add an equation that does not constrain the model in any way. It simply defines the auxiliary variable pop to be the sum of our decision variables. This allows us to express the objective easily and, possibly, to help display the solution.

The objective function is on line 11, a translation of (1.4). The function choices are, unsurprisingly, either s.Maximize or s.Minimize with, for parameter, a linear expression in terms of the variables declared previously.

We used

```
s.Maximize(pop)
```

We could have written

```
s.Maximize(x[0]+x[1]+x[2])
```

We then call on the solver at line 12 to do its job. This is where all the computational work gets done, work that I will not describe. The interested reader can search for "simplex method" and "interior-point methods" to learn about the fascinating theory[11] behind the solution methods of linear optimization models. To understand the simplex method, one needs only high school algebra. To understand interior-point methods requires a somewhat more mathematical background.

For some models, solvers may complete their work in a fraction of a second; for others, it may take hours. Moreover, not all solvers will have the same runtime behavior. Model A may run faster than model B on solver X while it may be exactly the reverse on solver Y. One more advantage of using the OR-Tools library is that we can try out another solver by changing one line.

---

[11]See, for example, Alexander Schrijver, *Theory of Linear and Integer Programming* (Hoboken, NJ: Wiley, 1998).

We should, if this code were meant for production and the problem nontrivial, check the return value to ensure that the solver found an optimal solution. It may have aborted because of a model error, or because it ran out of time or memory, or for some other reason. But for this simple first example, we will forgo good engineering practice in the name of simplicity of exposition.

We return, on line 13, both the optimal objective function value held in variable pop and the optimal values of the decision variables (not all the associated object attributes carried by those variables).

On more complex models, we may post-process the decision variables to return something simpler and more meaningful to the caller. You will see a good example of this when we solve the shortest path problem in Chapter 4, Section 4.4. The general approach I encourage is to create models that can be used without any knowledge of the internals of OR-Tools. The modeler is responsible for the creation of the model, but once the model is created and validated, it should leave the hands of its creator for those of the domain expert who originally formulated the problem.

When the diligent reader executes Listing 1-2, she will observe a result similar to Table 1-3.

Listing 1-2. How to Execute the Coexistence Model

```
1  from __future__ import print_function
2  from coexistence import solve_coexistence
3  pop,x=solve_coexistence()
4  T=[['Specie', 'Count']]
5  for i in range(3):
6    T.append([['Toads','Salamanders','Caecilians'][i], x[i]])
7  T.append(['Total', pop])
8  for e in T:
9    print (e[0],e[1])
```

Table 1-3. *Solution to the Coexistence Problem*

Specie	Count
Toads	100.0
Salamanders	300.0
Caecilians	1000.0
Total	1400.0

Notice that you can look at the solution of Table 1-3 and see that it does indeed satisfy the constraints. By substituting the solution into (1.1)-(1.3), we obtain

$2(100.0) + 300.0 + 1000.0 = 1500$	$\leq 1500,$
$100.0 + 3(300.0) + 2(1000.0) = 3000$	$\leq 3000,$
$100.0 + 2(300.0) + 3(1000.0) = 3700$	$< 5000.$

Notice that the first two inequalities are satisfied with equality. In the jargon of optimization, such inequalities are *tight* or *active*. The last one is said to be *slack* or *inactive*. In a certain sense, we could delete it from the problem and nothing would change. (The reader can try this and other modifications. The code is available in the additional material under the name coexistence.py).

In summary, the steps to construct and run a model are the following and are shown in Figure 1-1:

- Formulate the question precisely.

- Define the decision variables by identifying what is required to answer the question.

- Possibly define auxiliary variables to help simplify the statements of constraints or of the objective function. They can also help in the analysis and the presentation of the solution.

- Translate each constraint into an algebraic equality or inequality involving directly the decision variables or indirectly through the auxiliary variables.

- Construct the objective function as some quantity that should be minimized or maximized.

- Run the model using an appropriate solver.

- Display the solution in an appropriate manner.

- Validate the results. Does the solution correctly satisfy the constraints? Is the solution meaningful and implementable? If so, declare that you are done; if not, consider the necessary modifications to the model.

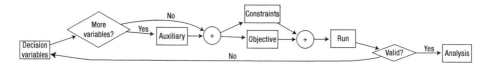

Figure 1-1. *The steps to construct and run a model*

The rest of this book will construct models of increasing complexity, illustrating and expanding the points above.

CHAPTER 2

Linear Continuous Models

At the dawn of optimization (the nineteen fifties), the state-of-the-art was defined by linear optimization models and the simplex method, the only reasonably efficient algorithm known at the time to solve such models. When I started studying this subject, one repeatedly heard from multiple sources that over 70% of the CPU cycles in the world were devoted to running various simplex codes. Surely an exaggeration, but it is indicative of the power of linear models. The world is not linear, but sometimes a linear approximation is **good enough**.

More precisely, I discuss here *linear continuous* models (though the usage is to call these models LPs for linear programs, implying the continuity properties). Linear continuous models are the simplest to write down and the simplest to solve. They have been the workhorse of optimizers since George Dantzig invented the simplex method to solve them. What characterizes them are three elements:

- All variables are continuous.

- All constraints are linear.

- The objective function is linear.

In detail, the decision variables (say $x_0, \dots, x_n$) can take on integral and fractional values. This is appropriate when the solution is measuring

© Serge Kruk 2018
S. Kruk, *Practical Python AI Projects*, https://doi.org/10.1007/978-1-4842-3423-5_2

amounts (for example, pounds of flour or tons of concrete). It is not appropriate when the solution is counting objects (as in people or politicians), unless one is looking only for an approximation.

The objective function is (or can be) parameterized by constant array c and expressed as

$$c_1 x_1 + c_2 x_2 + \ldots + c_n x_n$$

This limitation precludes objective functions with terms of the form

$$x_1^2, x_4^3, \sin(x), e^{x_3}, x_1 \cdot x_2, \ or \ |x|$$

among infinitely many others, although you will see later how to handle some of these non-linearities by model transformation.

Finally, the constraints are parameterized by matrix a_{ij}, array b, and can be stated as a set of relations, for $i \in \{1, \ldots, m\}$,

$$a_{i1} x_{i1} + a_{i2} x_{i2} + \ldots + a_{in} x_{in} \geq b_i, \text{ or } \leq b_i, \text{ or } = b_i$$

or some equivalent algebraic form.

In this chapter, we consider problems where the natural formulation is such a linear continuous model.

2.1 Mixing

The canonical linear programming example is the *diet problem*, one of the first optimization problems to be studied in the thirties and forties (twentieth century, not twenty-first[1]). The likely apocryphal origin of the problem is the US military's desire to meet the nutritional requirements of the field GIs while minimizing the cost of the food. One of the early researchers to study this problem was George Stigler. He made an

---

[1] I add this temporal precision on the odd chance that this text is still being read long after my body has maximized its entropy.

educated guess of the optimal solution to the linear program using a heuristic method. In the fall of 1947, Jack Laderman of the Mathematical Tables Project of the National Bureau of Standards (NBS, today NIST) undertook solving Stigler's model with the new simplex method. The linear model consisted of nine equations in 77 unknowns, a huge problem for the time. Some models in this book are orders of magnitude larger and will be solved in a minuscule fraction of the time it took the NBS people to solve the diet problem in 1947. The increase in efficiency is partly due to the hardware, but mostly due to the software.

A generic version of the problem is

> Given a list of food with some nutritional content, each with a cost, find the combination of food that will minimize cost and yet provide all the necessary nutrients.

Here is one simple version of this problem. The foods are F0, F1, F2, F3, etc. (Imagine them to be pizza, ramen noodles, cupcakes, chips, etc. or, if you are of a more health-conscious bent, tofu, green peas, quinoa, beets, etc.) The nutrients will be represented by N0, N1, N2, N3, etc. (Imagine them to be calories, protein, calcium, vitamin A, etc.). Each has a cost per serving. In addition, to avoid eating one food all week long, let's restrict the number of servings per week.

A randomly generated instance is given in Table 2-1.[2] Each row represents a food, with the nutritional content per serving followed by the acceptable range of servings of the food and its cost per serving. Ignore the last row and column for now. We will return to them after the model is constructed and solved. The two rows before last represent the allowed range of each nutrient.

---

[2]To encourage the reader to experiment, every model in this book is available in the additional material (https://github.com/sgkruk/Apress-AI), along with a random instance generator.

Table 2-1. *Example of Data and Solution for the Diet Problem*

	N0	N1	N2	N3	Min	Max	Cost	Solution
F0	606	563	665	23	7	17	9.06	17.0
F1	68	821	83	72	6	27	8.42	7.47
F2	28	70	916	56	1	36	9.47	6.11
F3	121	429	143	38	14	26	6.97	14.0
F4	60	179	818	46	9	35	4.77	35.0
Min	5764	28406	48157	1642				
Max	15446	76946	82057	6280				
Sol	14775	28406	48157	3413			539.37	

2.1.1 Constructing a Model

What would a solution be but a list of servings of each food? Therefore, the decision variables must be one per food, representing the number of servings. Let's name these variables $f_0, \ldots, f_n$. We will assume that it is acceptable to have fractional answers (i.e. one half serving is acceptable).

The objective is to minimize cost. We have one cost per food ($c_0, \ldots, c_n$). These are not variables, they are data. Therefore, what we want is to minimize the sum of all the products, $c_i \times f_i$. This leads to the objective function

$$\min \sum_i c_i f_i$$

Let's tackle the constraints. We have two sets: one indicating the range of acceptable servings of each food (assume that the minimum of food i is l_i and maximum is u_i) and one indicating the required nutrients range (minimum of nutrient j is a_j and maximum is b_j). The simpler constraint is related to the food. Since our decision variables indicate the number of servings of each food, we need only to box each serving count,

$$l_i \le f_i \le u_i \tag{2.1}$$

The constraint on nutrients is a bit more involved. Consider nutrient j. How much of it will be included in the diet? Each food i may have some of it, as indicated in Table 2-1. Let's call this amount N_{ji} (corresponding to the entry at the row of food i and the column of nutrient j). To get the total of this nutrient, we therefore need to sum over all foods the product of the food serving and the nutrient content. For each nutrient j,

$$a_j \le \sum_i N_{ji} f_i \le b_j$$

We are done with the theory. Let's translate this into an executable model general enough to solve all problems of this type (Listing 2-1). We will assume that the data is given in a two-dimensional array called N. It has the structure of Table 2-1 without the last column and row. Each row represents a food, except that the last two rows represent the minimum and maximum requirement of each nutrient, represented by the columns, with the last three representing the minimum, maximum, and the cost of each food serving.

Listing 2-1. Model for Minimal Cost Diet (diet problem.py)

```
1  def solve_diet(N):
2      s = newSolver('Diet')
3      nbF,nbN = len(N)-2, len(N[0])-3
4      FMin,FMax,FCost,NMin,NMax = nbN,nbN+1,nbN+2,nbF,nbF+1
5      f = [s.NumVar(N[i][FMin], N[i][FMax],'') for i in
         range(nbF)]
```

```
6    for j in range(nbN):
7        s.Add(N[NMin][j]<=s.Sum([f[i]*N[i][j] for i in
         range(nbF)]))
8        s.Add(s.Sum([f[i]*N[i][j] for i in
         range(nbF)])<=N[NMax][j])
9    s.Minimize(s.Sum([f[i]*N[i][FCost] for i in range(nbF)]))
10   rc = s.Solve()
11   return rc,ObjVal(s),SolVal(f)
```

The model uses the newSolver function to simplify the expression of the code[3] as the reader can see at Listing 2-2. These, and other simplifications, can be found in my_or_tools.py.

Listing 2-2. Utility Function to Create an Appropriate Solver Instance

```
from ortools.linear_solver import pywraplp
def newSolver(name,integer=False):
  return pywraplp.Solver(name,\
                    pywraplp.Solver.
                        CBC_MIXED_INTEGER_PROGRAMMING \
                    if integer else \
                    pywraplp.Solver.GLOP_LINEAR_PROGRAMMING)
```

To help the expression of the model, lines 3-4 give meaningful names to the row and column indices that we will use. In line 5, we define the decision variables, one per food, each taking values in the range $[l_i, u_i]$ as in equation (2-1). It would be correct to give a range of $[0, +\infty)$ and then add constraints to enforce the bounds. The solver would still find the same solution, but it is simpler and good practice to limit as much as possible

---

[3]Mostly to make the code fit a page, but also to hide some of the verbosity of the OR-Tools library. The authors chose, rightly in my opinion, meaningful but rather long names for their functions.

the range of decision variables. In complex models, it often dramatically improves the solution time.

The two-line loop starting on line 6 establishes the range on each nutrient as in (2.1.1). Lines 9 and following create the objective function, solve the problem, and return three numbers: the status of the solver (it should be zero), the optimal value, and the optimal solution. The dual role of the functions SolVal and ObjVal (seen in Listing 2-3) is to simplify the results returned to the caller and the code to read.

Listing 2-3. Utility Functions to Extract Values from the OR-Tools Objects

```
def SolVal(x):
  if type(x) is not list:
    return 0 if x is None \
      else x if isinstance(x,(int,float)) \
        else x.SolutionValue() if x.Integer() is False \
          else int(x.SolutionValue())
  elif type(x) is list:
    return [SolVal(e) for e in x]

def ObjVal(x):
  return x.Objective().Value()
```

The results from executing this model are shown in the last row and column of Table 2-1. The column indicates the number of servings of each food and the row indicates the amount of each nutrient that will be in the diet. The reader should notice that many of the food items and nutrient counts are at their minimum required values. This is expected of such a model since we are trying to minimize a linear cost function; the optimal solution should push towards the boundary of the constraints as much as possible.

The reader can experiment with this model. It is included in the additional material as `diet problem.py`, along with a generator of random diet problems and a routine to display the solution in a table format similar to Table 2-1.

2.1.2 Variations

There are a number of simple variations of this problem.

- Instead of minimizing cost, we could be given a profit to maximize.

    We could also not have either the minima or the maxima in either the foods or nutrients.

- It becomes more complex, and consequently interesting, when we have, in addition, requirements of the form "If food 2 is used, then we must have at least as much food 3 in the diet" or "Nutrient 3 must be included in at least twice the amount as nutrient 4."

    Let's consider some of these in detail. First, let's try "If food 2 is used, then food 3 must also be included in at least as many servings." The following inequality ensures the required result:

$$f_3 \geq f_2$$

    Notice that food 3 could still be included when food 2 is not, but that does not violated the requirement. And if food 2 is included, then we will have at least as many servings of food 3. It should be clear that the requirement could have been stated in reverse as "No more food 2 than food 3." The constraint is the same.

- A requirement on the nutrients, "Nutrient 3 must be included in at least twice the amount as nutrient 4," has a similar flavor but note that the amount of any given nutrient is spread among all food items. It may be fruitful to introduce auxiliary variables that will tally the nutrients, say n_j. We then add to the model one equality per nutrient,

$$n_j \leq \sum_i N_{ij} f_i$$

Note that these equations do not constrain the problem; their insertion is simply a helpful device to implement the requirement. We can now easily relate the nutrient content according to the new requirement as

$$n_3 \geq 2n_4$$

This we could have stated, had we not defined the variables n_i, as

$$\sum_i N_{i3} f_i \geq 2 \sum_i N_{i4} f_i$$

Defining the auxiliary variables n_j seems clearer. Moreover, displaying the total of each nutrient at the end might help with the analysis or the presentation of the solution.

- A similar requirement may occur to the reader, namely "If food (nutrient) 3 is used then food (nutrient) 4 must not be (and vice versa)." This may look like a simple variation to the above but it is decidedly **not** simple. If fact, it forces the modeler to use a different modeling technique. You will see how to implement such requirements in later chapters (see, for instance, Section 7.2 in Chapter 7). There are two valid approaches to modeling such requirements properly: integer programming and constraint programming. The reader is encouraged to spend some time trying to model such constraints to develop intuition into the difficulties. The key, and the reason that this is a beast of an entirely different ilk, is that the change is not uniquely quantitative (as much as, or twice the amount of) but is additionally qualitative: we transition between having an element and not having that element.

2.1.3 Structure of the Problems Under Consideration

Problems with the structure of the diet problem are generally known as *product mix* problems. They can be presented in various ways but if they can be fitted into the abstract Table 2-2 they can all be handled in the manner described in this section. Of course, it may be that some of the columns or rows are missing (no cost, or no price, or no maximum demand, etc.) That only simplifies the model.

Table 2-2. *Abstract Structure of Product Mix Problems*

		Components			Availabilities		Cost
		C_1	...	C_n	Min	Max	
	P_1	99	...	99	99	99	99
Products	...	...	...	...	...	...	...
	P_m	99	...	99	99	99	99
Demands	Min	99	...	99	99	99	99
	Max	99	...	99	99	99	99
Price		99	...	99	99	99	99

The decision variable indicates the amount of product needed and the constraints indicate availabilities of the raw material or, equivalently, capacities of the processing units as well as demand bounds. The objectives are often profits to maximize or costs to minimize or, simply, quantity to produce.

Here are a few instances to help the reader recognize the underlying structure. The reader is encouraged to marshal the problems into the format of Table 2-2 by inventing numbers.

- A factory is producing cement of various types. Each product is composed of the same elements, but in various quantities, and we have on hand a limited supply of each of these elements, each with a cost. To each final product is associated a profit. What is the best mix of product to produce to maximize profit?

- A Florida-based fruit company produces orange drinks, juices, and concentrates for various markets. The raw materials for all products are oranges, sugar, water, and time in various quantities, some positive and some negative (producing orange drinks requires water; producing concentrates generates water). Given certain availabilities, how much can the company produce to maximize profit?

- A toy manufacturer produces a number of different toys. Each is composed of a number of basic materials and, in addition, requires special processing (assembling, painting, boxing). The processing is done on specialized machines and has a duration. Since the manufacturer has limited supplies of materials and machines, which can only operate a certain number of hours per day, how many toys can be produced?

- A fertilizer company named Bush, Rove and Company (BR & Co.) has two products: a high phosphate blend and a low phosphate blend. They are produced by mixing different raw materials in various quantities.

  The company can procure, from its own subsidiaries, at most some amount of each raw material per day at a fixed internal cost. This cost includes labor, power, depreciation, delivery, bribes, etc. In addition, the mixing process incurs a certain cost per ton for each product.

  Both products are sold to a wholesaler, Fox Inc., at a fixed price. Moreover, the wholesaler has agreed to buy all the production BR & Co. can produce. How much of each fertilizer should it produce?

- Queequeg sells half-kilo bags of coffee in three blends, House, Special, and Gourmet, which sell at different prices per bag. Each blend is made up of Colombian, Cuban, and Kenyan coffee beans in various proportions. Queequeg has on hand some Colombian, Cuban, and Kenyan. How much of each blend should it bag to maximize revenues?

2.2 Blending

A second type of problem that readily admits a linear model is the *blending problem*. The classical example involves blending so-called raw or crude gasolines to achieve various refined products with specified octane value. For instance, let's assume that we are given crude gasolines R0, R1, ... , Rn, each with a certain octane rating, maximum availabilities in barrels, and a cost in dollars per barrels, as shown in Table 2-3.

Table 2-3. *Example of Raw Gasolines for the Blending Problem*

Gas	Octane	Availability	Cost
R0	99	782	55.34
R1	94	894	54.12
R2	84	631	53.68
R3	92	648	57.03
R4	87	956	54.81
R5	97	647	56.25
R6	81	689	57.55
R7	96	609	58.21

We are also given demands for multiple types of refined gasolines, (think Bronze, Silver, and Gold) with their own octane ratings. The demands are stated in minimum and maximum number of barrels along with their selling prices, as shown in Table 2-4.

Table 2-4. *Example of Refined Gasolines for the Blending Problem*

Gas	Octane	Min. Demand	Max. Demand	Price
F0	88	415	11707	61.97
F1	94	199	7761	62.04
F2	90	479	12596	61.99

We create the three types by mixing the appropriate raw gasolines together, assuming that the octane rating of a mix is a linear function of the volumes mixed. This is a crucial assumption: if we mix half and half of octane ratings 80 and 90, we get an octane rating of 85 because

$$\frac{1}{2}80 + \frac{1}{2}90 = 40 + 45 = 85$$

If we mix 40% of octane 80 and 60% of octane 90, we get

$$\frac{40}{100}80 + \frac{60}{100}90 = 32 + 54 = 86$$

This assumption is the key to the blending model.

Notice that there might be a number of ways to mix the raw gasolines together to get the required ratings. Our task is to construct a model that will tell us exactly how to mix the raw products to satisfy the demands and maximize the profits (understood as the difference between the total selling price of the finished products and the total cost of the raw gasolines).

2.2.1 Constructing a Model

What is the question to answer? Let's ask this question a number of times with increasing precision. A first stab is "How much of each type of refined gas to produce?" This is correct but is incomplete, since we need to know the composition of each refined gas, how much of each crude goes into each mix. A second stab is "How to mix the crude gas to produce the refined gas?" This is the right question, but is not yet in the proper form for an algebraic model. Imagine you are the manager of the refinery. On one side you have all these tanks filled with crude gas, and on the other side all empty tanks that will contain the refined gas. In between: miles of pipes with valves that you control. What you really want to know is which valves to open and by how much to have exactly the right mix. So the proper question is "How much of each crude gas goes into each refined gas?"

The key difference between the *mixing* problems of the preceding section and this *blending* problem is that previously we were told the exact composition of the products in terms of the material (in each food, the amount of each nutrient, for instance) while in the problem considered here, the composition of each product is one of the answers sought.

Since we need to know how much of raw i goes into refined j, we are led to a two-dimensional decision variable, say G_{ij} where i is the index of the crude gas and j is the index of the refined gas. For example, $G_{51} = 250$ will mean that there are 250 barrels of crude 5 going into the mix of refined 1. We understand here that the units will be barrels; it seems natural because the prices are per barrels. We probably should also introduce auxiliary variables to help us model and present the solution: the total of each crude gas (the sum of a row of G), say R_i, and the total of each refined gas (the sum of a column of G), say F_j. So we will have these non-constraining equations in the model:

$$R_i = \sum_j G_{ij} \quad \forall i \tag{2.2}$$

and

$$F_j = \sum_i G_{ij} \quad \forall j \tag{2.3}$$

Note that, by construction, $\sum_i R_i = \sum_j F_j$. That is, the total volume of crude used is equal to the total volume of refined products. We need not enforce this, though we need it. We can think of this as a "continuity" equation: it reflects that the refining process does not lose product along the way. This idea of continuity is a useful modeling idea. It will reappear in various guises throughout the models we develop.

Armed with these variables, we can now easily model the objective function. We are asked to maximize profits, hence the difference between total sales (given price p_j for refined gas j) and costs (given cost c_i for crude gas i),

$$\max \sum_j F_j p_j - \sum_i R_i c_i$$

The constraints come in multiple forms. The easy ones are, as in the mixing problems, constraints on the availability of each raw material. With our auxiliary variables, these are simple to express and can be included in the range of the defined variables or in a constraint

$$0 \le R_i \le u_i \quad \forall i$$

The constraints on the demand of refined gas (minimum and/or maximum) are just as simple:

$$a_j \le F_j \le b_j \quad \forall i$$

Notice how our auxiliary variables help write down constraints. Having only our decision variables, the constraints would have to be written with respect to column and row sums.

The only real complication of this problem refers to the octane rating. The key here is the assumption of linearity. To see how to model the octane requirement, let's imagine a simple case: say we mix 800 barrels of crude 1 with octane rating of 98 with 200 barrels of crude 2 with octane rating of 90. What is the resulting octane rating? Since we have a total of 1,000 barrels of refined,

$$\frac{800 \times 98 + 200 \times 90}{1000} = 96.4$$

So, in general, we need the fraction of each crude that goes into a mix times its octane rating. Assuming O_i as the octane rating of crude i and o_j the octane rating of refined j, this leads us to

$$\sum_i O_i G_{ij} = F_j o_j \quad \forall j \tag{2.4}$$

We now have an algebraic linear model. Let's translate it into executable code, as in Listing 2-4. We will assume that the data are entered in two-dimensional arrays, exactly as in Tables 2-3 and 2-4, except for the first columns, added for reference only.

Listing 2-4. Gasoline Blending Model (gas blend.py)

```
1  def solve_gas(C, D):
2    s = newSolver('Gas blending problem')
3    nR,nF = len(C),len(D)
4    Roc,Rmax,Rcost = 0,1,2
5    Foc,Fmin,Fmax,Fprice = 0,1,2,3
6    G = [[s.NumVar(0.0,10000,'')
7           for j in range(nF)] for i in range(nR)]
8    R = [s.NumVar(0,C[i][Rmax],'') for i in range(nR)]
9    F = [s.NumVar(D[j][Fmin],D[j][Fmax],'') for j in
     range(nF)]
10   for i in range(nR):
11     s.Add(R[i] == sum(G[i][j] for j in range(nF)))
12   for j in range(nF):
13     s.Add(F[j] == sum(G[i][j] for i in range(nR)))
14   for j in range(nF):
15     s.Add(F[j]*D[j][Foc] ==
16           s.Sum([G[i][j]*C[i][Roc] for i in range(nR)]))
17   Cost = s.Sum(R[i]*C[i][Rcost] for i in range(nR))
18   Price = s.Sum(F[j]*D[j][Fprice] for j in range(nF))
19   s.Maximize(Price - Cost)
20   rc = s.Solve()
21   return rc,ObjVal(s),SolVal(G)
```

At lines 3-5 we declare some constants to access the appropriate rows and columns of the data. The constraints on the range of each variable are entered not as constraints, but rather as a range on the corresponding variables. The equations (2.2)-(2.3) are seen on the four lines starting at 10.

The blending equations are created on the loop of line 14. Note that since the goal is to achieve a certain octane level, we might replace the equality with an inequality, indicating that the refined product has *at least* the required octane level. This relaxes the problem a little and allows optimization over a larger space. This might be required if, for example, we did not have sufficient low octane crude gasolines available.

The objective function (three lines starting at 17) maximizes the difference between the selling price of the refined product and the cost of the crude gas used.

Executing this model with the data above produces Table 2-5 where the bottom right number is the profit: the difference between the sum of the row *Price* and the column *Cost*.

Table 2-5. *Complete Solution to the Blending Problem*

	F0	F1	F2	Barrels	Cost
R0	542.5		239.5	782.0	43275.88
R1		894.0		894.0	48383.28
R2	631.0			631.0	33872.08
R3		648.0		648.0	36955.44
R4	704.41	251.59		956.0	52398.36
R5		647.0		647.0	36393.75
R6	449.5		239.5	689.0	39651.95
R7	50.93	558.07		609.0	35449.89
Barrels	2378.33	2998.67	479.0		
Price	147385.32	186037.28	29693.21		36735.18

2.2.2 Variations

While blending problems can be presented in various ways, they can all be handled in the manner above. The decision variables should be two-dimensional: the sum in one dimension and the other indicating total input material used and total output material produced. Finally, in addition to the capacity and demand constraints, there should be at least one blending constraint satisfying a linearity assumption.

One interesting variation is that we might be asked to achieve more than one characteristic. For instance, in addition to an octane level, we might also be given a certain concentration of sulfur in each of the crude and asked to keep the refined gas below a certain sulfur threshold. In this case, the octane equation (2.4) will almost certainly need to be replaced by an inequality, ensuring a minimum octane level, and another similar inequality will ensure a maximum sulfur lever. Assuming S_i as the sulfur level of crude i and s_j as the sulfur level of refined j, we get

$$\sum_i O_i G_{ij} \leq F_j o_j \quad \forall j$$

and

$$\sum_i S_i G_{ij} \geq F_j s_j \quad \forall j$$

The reason for the inequalities is that it is unlikely for the problem to have any feasible solution with exactly the specified octane and sulfur levels. The reader might try to modify Listing 2-4 to verify this.

To help the reader recognize the underlying structure of blending problems, the following is an instance we will revisit soon, with additional complexities.

A very popular ingredient in junk food is manufactured by refining and blending various oils together. The oils come in five flavors (O1 to O5) and measures of "hardness" as given in Table 2-6, where cost is in dollars per tons and the hardness is measured in the appropriate unit.

Table 2-6. *Caption Needed*

	O1	O2	O3	O4	O5
Cost	110	120	130	110	115
Hardness	8.8	6.1	2.0	4.2	5.0

The oils O1 and O2 can be refined at production facility A, which has a capacity of 200 tons per month, while O3, O4, and O5 can be refined at production facility B, which has a capacity of 250 tons per month. There is no loss of weight during the refining process and you can ignore the cost of the process.

The final product is obtained by mixing various amounts of the five oils. It has a hardness restriction. Measured in the same unit as given in the table, it must lie between 3 and 6 units. It is assumed that hardness blends linearly. That is, if we mix 10 tons of oil O1 with 20 tons of oil O2, the blend will have a hardness rating of

$$(10 \times 8.8 + 20 \times 6.1) / (10 + 20)$$

The final product sells for $150 per ton. How should the oils be refined and blended to maximize profit?

2.3 Project Management

Project management, as is usually understood in the context of optimization, refers to a set T of tasks, each with two properties:

- A duration
- A subset of T (possibly empty) of preceding tasks

The classic example is house construction: tasks include finding location, drawing plans, getting permits, breaking ground, laying foundations, building walls, installing plumbing, bribing inspectors, etc. Crucially, some tasks must be done before others: you cannot build the roof until you raise the walls. The main question under consideration: "When should each task start to minimize the total project completion time?" That is, when do we start each task to have the house entirely built in the shortest time possible? Also, if one task falls behind schedule, what is the impact on all the ulterior tasks and how do we reschedule them?

Table 2-7 is an instance of such a project and I will use to illustrate a solution technique.

Table 2-7. *Example of Project Management Tasks*

Task	Duration	Preceding Tasks
0	3	{ }
1	6	{ 0 }
2	3	{ }
3	2	{ 2 }
4	2	{ 1 2 3 }
5	7	{ }
6	7	{ 0 1 }
7	5	{ 6 }
8	2	{ 1 3 7 }
9	7	{ 1 7 }
10	4	{ 7 }
11	5	{ 0 }

2.3.1 Constructing a Model

What we need to decide in this instance is how early to start each task, respecting precedence, to minimize the total completion time. This suggests, as a decision variable, the starting time of each task in the same units as the given durations. Let's assume a set T of tasks (corresponding to the first column of Table 2-7) to declare our decision variables as

$$0 \le t_i \quad \forall i \in T$$

To ensure that precedence requirements are met, let's assume that we have, in addition to duration D_i (corresponding to the second column of Table 2-7), subsets $T_i \subset T$ of preceding tasks for each task i (corresponding to the third column of Table 2-7). Then we need to lower bound the starting times by

$$t_j + D_j \le t_i \quad \forall j \in T_i; \forall i \in T$$

The objective is to minimize the project completion time. This time would be the starting time of the last task plus its duration if the tasks were all done sequentially. But they are likely not; we might be doing as many tasks in parallel as possible. Then how do we find the completion time if we do not know the last task, or if there is no single "last" task?

Let's introduce another variable, t. We will constrain this t to be larger than, for each task, its starting time plus its duration. It will therefore be larger than the completion time. And if we add the objective min t to the set of constraints

$$t_i + D_i \le t \quad \forall i \in T$$

then t will, at optimality, be the completion time, a condition that will hold no matter how many tasks we do in parallel.

This is translated into an executable model in Listing 2-5 where we assume that the data is given to us in table D with the same structure as

Table 2-7: each row has a task identifier, a duration, and a set, possibly empty of preceding tasks.

Listing 2-5. Project Management Model (`project management.py`)

```
1   def solve_model(D):
2       s = newSolver('Project management')
3       n = len(D)
4       max = sum(D[i][1] for i in range(n))
5       t = [s.NumVar(0,max,'t[%i]' % i) for i in range(n)]
6       Total = s.NumVar(0,max,'Total')
7       for i in range(n):
8           s.Add(t[i]+D[i][1] <= Total)
9           for j in D[i][2]:
10              s.Add(t[j]+D[j][1] <= t[i])
11      s.Minimize(Total)
12      rc = s.Solve()
13      return rc, SolVal(Total),SolVal(t)
```

Line 4 computes a valid upper bound on the times by adding all the durations. This is clearly an overestimate but is fine to use in the declaration of the decision variables at line 5. We declare the total completion time variable at line 6, which we use as an upper bound on all starting times plus duration at line 8. Finally, we add the precedence bounds at line 10. The results appear in Table 2-8 and, graphically, in Figure 2-1. Note that the last ending time is the total project completion time.

Table 2-8. *One Optimal Solution to the Project Management Problem*

Task	0	1	2	3	4	5	6	7	8	9	10	11
Start	0	3	0	3	9	0	9	16	26	21	24	23
End	3	9	3	5	11	7	16	21	28	28	28	28

Note that all tasks could have started at any time after their required tasks have ended. And in fact, depending on the solver used, the solution might look rather different. You can see an example of an alternate solution in Table 2-9. This situation of multiple optimal solutions offers us, as modelers, an opportunity to improve the model. In this particular case, it might be useful to start all tasks as early as possible. This will not affect the total completion time but might make the project more practical and less prone to delays if some tasks' duration were poorly estimated.

Table 2-9. *An Alternate Optimal Solution to the Project Management Problem*

Task	0	1	2	3	4	5	6	7	8	9	10	11
Start	0	3	0	3	9	0	9	16	21	21	21	3
End	3	9	3	5	11	7	16	21	23	28	25	8

Note finally that by looking at the graphical representation, it is clear that the subset of tasks 0,2,1,6,7,9 is critical in the sense that if any of them are delayed, the project completion time is delayed. On small projects, such a graphical representation is sufficient to identify the critical tasks. On larger projects, it might be profitable to identify these tasks programmatically. You will see one way to compute critical paths in Section 4.4.3 in Chapter 4 when I discuss longest paths.

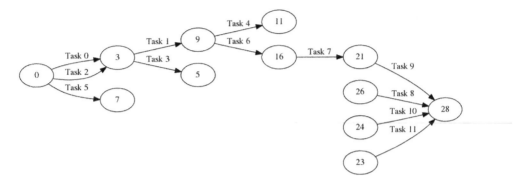

Figure 2-1. *Graphical representation of example solution (nodes are times)*

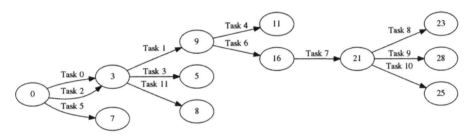

Figure 2-2. *Graphical representation of alternate solution*

2.3.2 Variations

I displayed two possible solutions to the problem in Figures 2-1 and 2-2. The alternate might be preferable for practical reason. How can we ensure that, among all solutions that minimize total completion time, we choose a solution that starts all tasks as early as possible? One way is to minimize the sum of starting times.

That is, replace the objective function by

```
s.Minimize(sum(t[i] for i in range(n)))
```

In cases like this, optimizers talk of *multiple objectives*. In general, these might be independent, or worse, contradictory. But in our project management situation, the objectives (minimizing completion time and starting all tasks as early as possible) are coherent. Note that the new optimal value of the model is neither interesting nor useful. We need to inspect the Total variable to give us the completion time.

2.3.2.1 Minimax Problems

The technique we used for the project management can be used more generally whenever we face a *minimax* problem. This is a problem where we want to minimize the maximum of some set of functions. For example, let's assume we want to find the optimal x for

$$\min_x \max_{i \in T} \sum_j a_{i,j} x_j$$

This is handled by introducing a new variable, say t, along with the objective

$$\min \ t$$

and the constraints

$$\sum_j a_{i,j} x_j \leq t \qquad \forall i \in T$$

The corresponding *maximin* problem is handled the similarly. Note that the related *maximax* and *minimin* are considerably more difficult to handle. We will revisit those in a later section (see section 7.2.4 in Chapter 7.)

2.3.2.2 Absolute Value Problems

Essentially the same approach can also be used for some non-linear functions, for instance, those involving absolute values. Say we seek the optimal x for

$$\min_x \left| \sum_j c_j x_j \right|$$

Since the absolute value function is defined as

$$|z| = \begin{cases} z & z \geq 0 \\ -z & z < 0 \end{cases}$$

we can use the same min t objective along with constraints

$$\sum_j c_j x_j \leq t \qquad \forall i \in T$$

and

$$-\left(\sum_j c_j x_j\right) \leq t \qquad \forall i \in T$$

I will illustrate some non-trivial applications of this technique in Section 3.2 in Chapter 3.

2.4 Multi-Stage Models

In life, decisions at one stage often influence decisions at a later stage. The same holds for more pedestrian situations. For instance, consider a warehouse: what it contains at the end of a month surely should influence what is ordered at the beginning of the following month.

In a certain sense there is little that is new in these multi-stage models except that we have to be careful to properly set up the continuity from one stage to the next.

To illustrate, let's revisit the blending problem. To multiple targets, prices, and costs, we will add a planning horizon of many months. This will exemplify the stages. This problem will require all the tricks and techniques you have seen so far (and then some). It forms a comprehensive review of the current chapter.

2.4.1 Problem Instance

Soap is manufactured by refining and blending various oils together. The oils come in various flavors (apricot, avocado, canola, coconut, etc.) and each oil contains multiple fatty acids (lauric, linoleic, oleic, etc.) in various proportions. For example, see Table 2-10.

Table 2-10. *Example of Oils (Oi) with Their Acid Content (Aj)*

	A0	A1	A2	A3	A4	A5	A6
00	36	20	33	6	4		1
01		68	13			8	11
02		6		66	16	5	7
03		32				14	54
04			49	3	39	7	2
05	45		40		15		
06						28	72
07	36	55					9
08	12	48	34		4	2	

According to the properties of the soap one is creating (cleaning power, lather production, dryness of the skin, etc.) one targets the final proportions of the fatty acids to be in certain ranges by blending the oils appropriately. For instance, we will target our soap to have acid contents in the ranges of Table 2-11.

Table 2-11. *Fatty Acid Content Targets*

	A0	A1	A2	A3	A4	A5	A6
Min	13.3	23.2	17.8	3.7	4.6	8.8	23.6
Max	14.6	25.7	19.7	4.1	5.0	9.7	26.1

Here is an additional twist, relative to periods. We will be planning for a certain number of months. Each oil may be purchased for immediate delivery or bought on the futures market for delivery in a later month. The price of each oil in each of the months is given in Table 2-12 in dollars per ton.

Table 2-12. *Cost of Oils in Dollars per Ton Over the Planning Horizon*

	Month 0	Month 1	Month 2	Month 3	Month 4
00	118	128	182	182	192
01	161	152	149	156	174
02	129	191	118	198	147
03	103	110	167	191	108
04	102	133	179	119	140
05	127	100	110	135	163
06	171	166	191	159	164
07	171	131	200	113	191
08	147	123	135	156	116

It is possible to store up to 1,000 tons of oil for later use (any combination of oils) but there is a holding cost of $5 per ton per month. Finally, we must satisfy a demand of 5,000 tons of soap per month. This demand drives the model.

At the beginning of the planning horizon, we have some oils in inventory, as illustrated in Table 2-13. How should the oils be refined and blended every month to minimize cost?

Table 2-13. *Initial Inventory in Tons*

Oil	Held
00	15
01	52
02	193
03	152
04	70
05	141
06	43
07	25
08	89

2.4.2 Constructing a Model

2.4.2.1 Decision variables

The question to answer is "How should the various oils be blended every month?" This means we need to identify how much of each oil goes into the final blend during each month. This is a good start but it is clearly not enough. For instance, we can blend from oil we buy and from oil we have in inventory.

So we need to distinguish these two quantities. Moreover, we may decide to buy for storage (because the prices are about to go up) so we also need to know how much we can store. This suggests at least three decision variables for each oil ($O = \{0, 1, 2, \dots, n_o\}$ will be the set of oils), and for each month ($M = \{0, 1, 2, \dots, n_m\}$ is the set of months)

$x_{i,j} \geq 0 \ \forall i \in O, \forall j \in M$	Buy
$y_{i,j} \geq 0 \ \forall i \in O, \forall j \in M$	Blend
$z_{i,j} \geq 0 \ \forall i \in O, \forall j \in M$	Hold

The interpretation is $x_{i,j}$ will be the number of tons of oil i bought during month j; $y_{i,j}$ will be the number of tons blended into our soap; and $z_{i,j}$ is the number of tons held at the beginning of the month. Note that we have a choice here to have the variable represent the amount at the beginning or at the end of the period. Either is acceptable but it must be clear in the model which one is chosen because it affects the constraints. A typical mistake in a multi-period model is to have some constraints assume that a variable represents a quantity at the start of the period while some other constraints assume the end. The model may run, but the solution will be nonsensical. Since we are given quantities in storage at the beginning of the planning period, having a variable represent the quantity held at beginning means that we can easily initialize it with the given data.

We probably will need to know how much soap we are producing each month. This is not, strictly speaking, essential to the problem as formulated, but it may make the presentation of the solution and maybe the formulation of some constraints much simpler. As usual, it helps to introduce auxiliary variables to clear up some statements. To tally the total production per month,

$$t_j \qquad \forall j \in M$$

2.4.2.2 Constraints

Let's tackle the continuity constraints. We need to specify for each oil and for each month (but the last) how the inventory fluctuates, so

$$z_{i,j} + x_{i,j} - y_{i,j} = z_{i,j+1} \qquad \forall i \in O, \forall j \in M \setminus \{n_m\} \tag{2.5}$$

In words, this is what is held at the beginning of the month plus what we buy minus what we blend forms the new inventory.

We have a minimum and a maximum storage capacity at each month of the total amount of oil, or

$$C_{min} \leq \sum_i z_{i,j} \leq C_{max} \quad \forall j \in M$$

Now comes the blending constraint, or rather constraints, since we need to target a number of fatty acids. To help the formulation, let's extract the total production,

$$t_j = \sum_i y_{i,j} \quad \forall j \in M$$

Let's assume that for each acid $k \in A$ we have a target range $[l_k, u_k]$ and that each oil $i \in O$, a percentage p_{ik} of the required acid (Table 2-10). Since the final product for each acid must fall in a certain range, we should have two constraints: one for the low end and one for the high end of the interval. That is,

$$\sum_i y_{i,j} p_{i,k} \geq l_k t_j \qquad \forall k \in A, \forall j \in M \tag{2.6}$$

and

$$\sum_i y_{i,j} p_{i,k} \leq u_k t_j \qquad \forall k \in A, \forall j \in M \tag{2.7}$$

These constraints could be written without the production variables t_j but would be more cumbersome and difficult to read.

Finally, we need to satisfy demand. This is simple, assuming a demand of D_j at each month j,

$$t_j \geq D_j \quad \forall j \in M$$

2.4.2.3 Objective Function

We are told that the objective is to minimize costs, comprised of the varying oil costs at each month plus the fixed storage cost of the oils we keep in inventory. Therefore,

$$\sum_i \sum_j x_{i,j} P_{i,j} + \sum_i \sum_j z_{i,j} p$$

This type of objective (fixed plus variable cost) appears regularly in business-type problems. You will see this again when considering facility location to service customer demands. The decision to build incurs a fixed cost. The servicing of the various customers is a variable cost.

2.4.2.4 Executable Model

Let's now translate this into executable code as shown in Listing 2-6. There is a fair amount of data to pass in. Let's assume arrays `Part` as in Table 2-10, `Target` as in Table 2-11, `Cost` as in Table 2-12, and `Inventory` as in Table 2-13 in addition to three parameters: `D` in tons for the demand, `SC` in dollars per ton for the storage cost, and `SL` in tons for the minimum and maximum to hold in inventory.

From line 5 to line 11 we declare variables but only the first three are true decision variables. All the others are artificially introduced either to help us state the constraints (for 8 and 11) or to help us display some details of the resulting solutions. They will not affect the running time of the solver in any appreciable manner but will make our life easier.

At line 12 we set the `Hold` variable to contain what is known to be in the inventory at the start of the planning period.

The large loop starting at line 14 will set all the constraints since they have the identical structure for each month and we have declared our variables to be arrays indexed by the month.

Line 15 sets the artificial variable Prod to be the sum of the blended oils. This is not really a constraint, but rather a simplifying trick. If we repeat some calculations in a model as here,

```
sum(Blnd[i][j] for i in range(nO)
```

we should consider introducing an artificial variable. Assuming a decent solver, it will cost nothing and is likely to help. One of the principles of programming (and modeling) is "Do not repeat yourself."

We use this Prod variable immediately after, at line 16 to ensure that we satisfy the demand. If this demand is a scalar, we set it identically for each month, but it could be an array indexed by the month.

The code starting with the if on line 17 implements the continuity requirement we described in equation (2.5). We ensure that what we buy and what we have on hand at the beginning of the month equals what we blend and what we store for the next month. The conditional is to avoid setting a constraint on a month past the planning horizon.

Lines 20 and 21 ensure the bounds on the oils we keep in inventory.

The loop starting at line 22 first defines our auxiliary Acid variable to ease the formulation of the blending constraints stated on the following two lines, which correspond to equations (2.6) and (2.7). Acid, indexed by the ordinal of the fatty acid k and of the month j under consideration, is summed over all oils of the quantity blended with the oil's percentage of acid k. This quantity, divided by the total blended, will be the percentage that must fall within the required range.

Finally, the four lines starting at 26 set the artificial variables that will hold the costs of purchasing and holding at each period and then sum them to construct the objective function which we will minimize.

Listing 2-6. Multi-Period Blending Model (blend multi.py)

```
1   def solve_model(Part,Target,Cost,Inventory,D,SC,SL):
2       s = newSolver('Multi-period soap blending problem')
3       Oils= range(len(Part))
```

```
4    Periods, Acids = range(len(Cost[0])), range(len(Part[0]))
5    Buy = [[s.NumVar(0,D,'') for _ in Periods] for _ in Oils]
6    Blnd = [[s.NumVar(0,D,'') for _ in Periods] for _ in Oils]
7    Hold = [[s.NumVar(0,D,'') for _ in Periods] for _ in Oils]
8    Prod = [s.NumVar(0,D,'') for _ in Periods]
9    CostP= [s.NumVar(0,D*1000,'') for _ in Periods]
10   CostS= [s.NumVar(0,D*1000,'') for _ in Periods]
11   Acid = [[s.NumVar(0,D*D,'') for _ in Periods] for _
     in Acids]
12   for i in Oils:
13     s.Add(Hold[i][0] == Inventory[i][0])
14   for j in Periods:
15     s.Add(Prod[j] == sum(Blnd[i][j] for i in Oils))
16     s.Add(Prod[j] >= D)
17     if j < Periods[-1]:
18       for i in Oils:
19         s.Add(Hold[i][j]+Buy[i][j]-Blnd[i][j] == Hold[i]
           [j+1])
20     s.Add(sum(Hold[i][j] for i in Oils) >= SL[0])
21     s.Add(sum(Hold[i][j] for i in Oils) <= SL[1])
22     for k in Acids:
23       s.Add(Acid[k][j]==sum(Blnd[i][j]*Part[i][k] for i in
         Oils))
24       s.Add(Acid[k][j] >= Target[0][k] * Prod[j])
25       s.Add(Acid[k][j] <= Target[1][k] * Prod[j])
26     s.Add(CostP[j] == sum(Buy[i][j] * Cost[i][j] for i in Oils))
27     s.Add(CostS[j] == sum(Hold[i][j] * SC for i in Oils))
28   Cost_product = s.Sum(CostP[j] for j in Periods)
29   Cost_storage = s.Sum(CostS[j] for j in Periods)
30   s.Minimize(Cost_product+Cost_storage)
```

```
31  rc = s.Solve()
32  B,L,H,A = SolVal(Buy),SolVal(Blnd),SolVal(Hold),SolVal(Acid)
33  CP,CS,P = SolVal(CostP),SolVal(CostS),SolVal(Prod)
34  return rc,ObjVal(s),B,L,H,P,A,CP,CS
```

Since this model is of a certain complexity, the caller should examine the return code of the solver. It needs to be zero for the solution to be optimal. The most frequent non-zero return status will be for infeasibility. This may occur for a number of reasons, the most likely of which is that there is no combination of oil that will achieve our target fatty acid content.

The results of a run with all the above data is displayed in Table 2-14. It displays everything we need to know. The first set of lines, to be sent to Purchasing, specify how much of each oil to buy per month. The next set of lines, to be sent to Manufacturing, describe the exact recipe of the blending to do each month. Notice that the soap is created from different oils in each month to achieve the minimal cost. The next set of lines, to be sent to the Bean Counters, describes the inventory, the product costs, and the storage costs at each month. And finally, we can send to Quality Control the last set of lines, indicating the actual percentages of fatty acids achieved by the blending recipe.

The main point of this model is to present the complexity of real models along with some tricks on managing this complexity at the model level. A second point is to highlight some of the advantages of modeling in Python instead of in specialized modeling languages.

2.4.3 Variations

There are an infinite number of variations of such a complex model.

- The demand could vary at each month, as shown in Table 2-14.

Table 2-14. *Multi-Period Blending Results*

Buy qty	Month 0	Month 1	Month 2	Month 3	Month 4
00	1935.7	0.0	0.0	0.0	0.0
01	480.7	0.0	274.6	0.0	0.0
02	192.4	0.0	545.9	0.0	0.0
03	2835.0	1553.3	0.0	0.0	0.0
04	293.7	0.0	0.0	136.8	0.0
05	0.0	966.7	1611.3	0.0	0.0
06	482.6	1011.5	275.1	1517.9	0.0
07	0.0	0.0	0.0	1247.9	0.0
08	0.0	1468.5	2293.1	597.4	0.0
Blend qty	**Month 0**	**Month 1**	**Month 2**	**Month 3**	**Month 4**
00	1683.6	117.7	149.4	0.0	2034.4
01	532.7	0.0	274.6	0.0	919.5
02	113.3	272.1	269.3	276.6	105.6
03	1551.3	1465.1	1524.0	0.0	382.6
04	363.7	0.0	0.0	136.8	392.7
05	141.0	966.7	1051.8	559.5	0.0
06	525.6	684.9	601.7	1517.9	1165.2
07	0.0	25.0	0.0	747.9	0.0
08	89.0	1468.5	1129.2	1761.3	0.0

(*continued*)

Table 2-14. (*continued*)

Hold qty	Month 0	Month 1	Month 2	Month 3	Month 4
00	15.0	267.2	149.4	0.0	0.0
01	52.0	0.0	0.0	0.0	0.0
02	193.0	272.1	0.0	276.6	0.0
03	152.0	1435.7	1524.0	0.0	0.0
04	70.0	0.0	0.0	0.0	0.0
05	141.0	0.0	0.0	559.5	0.0
06	43.0	0.0	326.6	0.0	0.0
07	25.0	25.0	0.0	0.0	500.0
08	89.0	0.0	0.0	1163.9	0.0
Prod qty	5000.0	5000.0	5000.0	5000.0	5000.0
P. Cost	$735098.96	$616064.04	$644688.93	$491829.66	$0.00
S. Cost	$3900.00	$10000.00	$10000.00	$10000.00	$2500.00
Acid %	Month 0	Month 1	Month 2	Month 3	Month 4
A0	13.6	13.3	13.3	14.6	14.6
A1	24.9	24.5	25.2	25.5	23.2
A2	17.8	18.5	17.8	17.8	19.7
A3	3.7	3.7	3.7	3.7	4.1
A4	5.0	5.0	5.0	5.0	5.0
A5	8.8	8.8	8.8	9.7	9.7
A6	26.1	26.1	26.1	23.6	23.6
Total	100.0	100.0	100.0	100.0	100.0

- Instead of satisfying some demand, we may be asked maximize profit. In this case, we need to know the price of the final product, which of course may change at each month.

- The inventory levels may be stated in terms of each oil instead of aggregate quantities.

- There may be uncertainty in the fatty acid content of certain oils.

2.5 Pattern Classification

Classification is currently one of the most successful applications of software to tasks that were, not so long ago, the privilege of the human intellect. For instance, software decides if an email is legitimate or spam, whether a biopsied cell is malignant or benign, and whether the company should offer you an interview or let your re´sume´ rot in the great bit bucket in the sky.

Let's look at one of the first effective techniques for the binary classification of data. The example is contrived because I want to draw pictures to guide the intuition, but the code we will write is applicable in a wide variety of cases.

Let's imagine that we are trying to automate the classification of cells as malignant or benign based on two measures: the area and the perimeter. Those features are measured automatically from a picture of the cell under a microscope. The process starts with a collection of such cells, divided by an expert into the two groups. These groups form what is known as the training set for our software. After we have "trained" our software, we will feed it new data, that has not been seen by an expert, and it will decide in which group the cell falls. That is, it will classify the cell as malignant or benign. This process is real and used in laboratories all over the world. The major simplification I am making here is that many more than two features are used in practice.

Let's consider as an example the cell features plotted in Figure 2-3 with perimeter on the x-axis and radius on the y-axis. We see that the two classes can be separated by a line. Our task is to discover that line. Of course, there are a number of valid lines but, as a first attempt, any line separating the two classes will do.

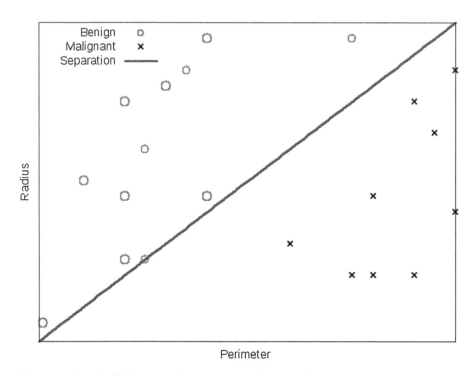

Figure 2-3. *Cell data and separation hyperplane*

2.5.1 Constructing a Model

Algebraically, a line is an equation of the form $a_1 x_1 + a_2 x_2 = b$ for some fixed coefficient a_1, a_2, a_0. Or, in dimension n, we call it a hyperplane and it has an equation of

$$\sum_{i=1}^{n} a_i x_i = a_0$$

What does it mean for a particular point x to be on one side or the other of the line? It means that either $a_1 x_1 + a_2 x_2 < a_0$ or $a_1 x_1 + a_2 x_2 > a_0$. These strict inequalities can be scaled to increase the gap by any amount. We can therefore simplify our task to identifying a vector a such that, for every point x' in class A, we have

$$\sum_i a_i x_i' \geq a_0 + 1$$

and that, for every point x'' in class B, we have

$$\sum_i a_i x_i'' \leq a_0 - 1$$

Let's introduce a positive variable for each of the data points, say y_i' for each point of class A and y_i'' for each of class B. Now the inequality $\sum_i a_i x_i \geq a_0 + 1$ can be enforced by requiring

$$y' \geq a_0 + 1 - \sum_i a_i x_i' \text{ and } y' \geq 0$$

and minimizing y' to zero. The algebra is symmetric for the points of class B. All in all, we are led to the following optimization problem:

$$\min \sum_{i \in A} y_i' + \sum_{i \in B} y_i''$$

$$\text{subject to } y' \geq a_0 + 1 - \sum_i a_i x_i',$$

$$y'' \geq \sum_i a_i x_i'' - a_0 + 1,$$

and

$$y', y'' \geq 0$$

One characteristic of this model is that if the optimal objective value is zero, we have a hyperplane correctly separating the training set into malignant cells and benign cells. But if the value is non-zero, it means that the set is not separable by a hyperplane and so more complex techniques are required.

2.5.2 Executable Model

Let's translate this into the executable model seen in Listing 2-7. It accepts two sets of data points with any number of features, classified by some expert into classes A and B. After defining the potential deviation from the hyperplane of sets A and B on lines 4 and 5, we define the variable that will hold the hyperplane on line 6. Note that we need this hyperplane later on, to do the classification of the unknown points. Note also that the coefficients could be restricted to be in any interval containing zero. It is simple to scale all coefficients of a plane to have its algebraic expression reside on whatever interval we choose, as long as it includes zero.

The constraints at lines 8 and 10 set up the offset of each point to the hyperplane which the objective function will attempt to minimize to zero.

Listing 2-7. Identification of the Classifying Hyperplane (features.py)

```
1   def solve_classification(A,B):
2     n,ma,mb=len(A[0]),len(A),len(B)
3     s = newSolver('Classification')
4     ya = [s.NumVar(0,99,'') for _ in range(ma)]
5     yb = [s.NumVar(0,99,'') for _ in range(mb)]
6     a = [s.NumVar(-99,99,'') for _ in range(n+1)]
7     for i in range(ma):
8       s.Add(ya[i] >= a[n]+1-s.Sum(a[j]*A[i][j] for j in
        range(n)))
9     for i in range(mb):
10      s.Add(yb[i] >= s.Sum(a[j]*B[i][j] for j in range(n))-
        a[n]+1 )
11    Agap = s.Sum(ya[i] for i in range(ma))
12    Bgap = s.Sum(yb[i] for i in range(mb))
13    s.Minimize(Agap+Bgap)
14    rc = s.Solve()
15    return rc,ObjVal(s),SolVal(a)
```

The reader might feel a little uncomfortable about this model and here is why, at least partly: This is a model where we do not care about the optimal value, but only whether it is zero or not. The decision variables (and we already discussed why this expression is such a misnomer) are not deciding anything. The set of y variables has no real interpretation other than it represents by how much a point violates a linear inequality. And finally, the only part of the solution we extract, the hyperplane, is not used yet. It will only be used later on, in a different program trying to classify a new point as belonging to class A or B. We have moved, with this model, to a higher abstract plane than ever before.

2.5.2.1 Variations

There are at least three directions we can go from this model.

- The first is to add constraints to increase the quality of the returned hyperplane. For example, we could require that it not only separates the two sets, but that it is, in some sense, as far from one set as from the other. If the training set is well-chosen, this will ensure that we minimize erroneous classifications later on. This is known as maximizing the margin and we will tackle this problem in a later chapter.

- The second direction to pursue is what to do when the optimal value is not zero; that is, when the two sets are not separable by a hyperplane. They may be separable by a nonlinear curve. This question is complex and multiple approaches have been tried, but most rely on knowing something additional about the data. We will not consider it.

- The final improvement would be to consider classification into multiple classes. We will consider this in a later chapter.

CHAPTER 3

Hidden Linear Continuous Models

In this chapter we do violence to some problems to reveal their inner structure. The focus is on problems which, at first glance, may not seem to be of the continuous linear variety yet can be marshalled into that form with a handful of creative alterations. The key is to ensure a one-to-one correspondence between the original and the altered problems so that we can retrieve a solution to the original from a solution to the alteration.

The main reason for massaging problems in this way is that continuous linear solvers have become so fast that they can handle models with hundreds of thousands of variables and constraints. Therefore, if a problem can be modeled in that manner, there is little practical limit on the instance size that can be solved. As you will see later, this is not the case with more complex models. In fact, we can write models with a few dozen variables that no current solver can solve in a reasonable time.

The main obstacles encountered in this chapter are non-linearities of one kind or another, but with the advantageous restriction that the functions be considered convex. A convex function[1] is one that sits "above"

---

[1] All research mathematicians agree on the labels "convex" and its opposite, "concave," but textbook authors for high schools in the US, ignoring thousands of papers, journals, and research monographs, insist on "concave up" and "concave down."

© Serge Kruk 2018
S. Kruk, *Practical Python AI Projects*, https://doi.org/10.1007/978-1-4842-3423-5_3

any valid linear approximation to it. In one dimension, algebraically, f is convex at point $x0$ if

$$f(x_0 + h) \geq f(x_0) + f'(x_0)h$$

Geometrically, it looks like Figure 3-1, with a first-order approximation of $f(x) = x^2$ at $x_0 = 4$. Convexity will be the Trojan horse used to beat the non-linearity into submission.

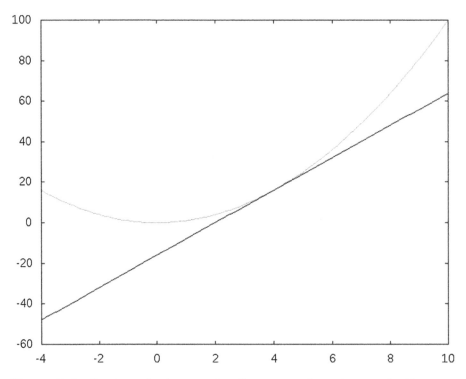

Figure 3-1. *Prototypical example of a convex function and a linear approximation*

3.1 Piecewise Linear

We consider here broken-up linear functions. In the traditional parlance, they are *piecewise linear*. As such, the linear programming solvers we have used up to now (GLPK, GLOP, CLP) cannot handle them directly, but a little coding on our part will morph them into a standard form that all solvers can handle. This is one of the good reasons to code models in Python instead of the specialized modeling languages.

As first example, to illustrate the technique without any side issue that might hide the essence, let's consider a piecewise function defined as

$$f(x) = \begin{cases} C_1 x & 0 \le x \le B_1, \\ C_1 B_1 + C_2 (x - B_1) & B_1 \le x \le B_2, \\ C_1 B_1 + C_2 (B_2 - B_1) + C_3 (x - B_2) & B_2 \le x \le B_3, \\ \cdots & \end{cases} \tag{3.1}$$

We can think of this function as a shipping cost function with additional penalties for weights; that is, the more product we ship, the more expensive each unit is. Table 3-1 (and Figure 3-2) represent an instance of this simple function where the first two columns bracket the quantities $(B_i, B_{i+1}$ for which the third column is the unit cost, c_i.

Table 3-1. *Example of Piecewise Function*

(From	To]	Unit Cost	(Total cost	Total cost]
0	148	24	0	3552
148	310	28	3552	8088
310	501	32	8088	14200
501	617	34	14200	18144
617	762	36	18144	23364
762	959	40	23364	31244

We will illustrate the approach by minimizing this function subject to a simple bound on the quantity.

3.1.1 Constructing a Model

What we need to decide in this problem is simply the quantity to produce. We can define a decision variable with bounds from 0 to the last quantity in the table as

$$x \in [0, B_n]$$

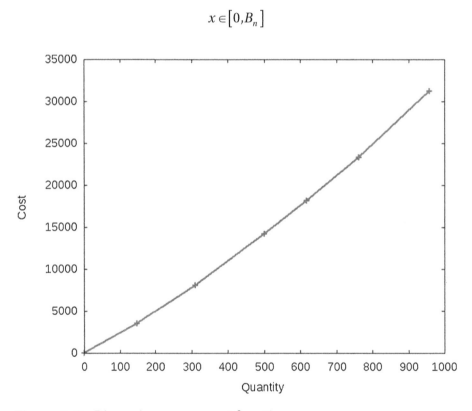

Figure 3-2. *Piecewise convex cost function*

But the quantity we decide on will affect the objective function, therefore we need to know in which bracket we end up, and where on that bracket. Here is the key trick: we introduce additional variables, one for each break-point in the function. Assuming that we have n brackets, conceptually we consider these variables as weights on the bracket boundaries, telling us where we are in the bracket. We want at most two consecutive variables to be non-zero, with their sum to be one. This will tell us where x lies and, consequently, what is the objective function value.

$$\delta_i \in [0,1] \; \forall i \in \{0,\ldots,n\}$$

For example, if $\delta_2 = \dfrac{1}{4}$ and $\delta_3 = \dfrac{3}{4}$ we know that we are in the third bracket, one quarter of the way and that $x = \delta_2 \, B_2 + \delta_3 \, B_3$.

3.1.1.1 Constraints

We will enforce that the δ sum be one, and, by the convex structure of the problem, at most two, adjacent δ will be non-zero. This will tell us which bracket and where in the bracket. To do this we add the constraint

$$\sum_i \delta_i = 1$$

We deduce the value of the decision variable by

$$x = \sum_i \delta_i B_i \tag{3.2}$$

Note that this x variable and its associated constraint play no role in the optimization model. For the solver, the δ is the real decision variable and the x is simply a translation into the language of the original problem. This is the key.

3.1.1.2 Objective

The objective function is linear within a bracket; therefore we sum over all brackets:

$$\min \sum_{i=1}^{n} \delta_i \sum_{j=1}^{i} \left(B_j - B_{j-1} \right) \times C_{j-1}$$

We must stress here that the transformation trick only works because of the structure of this objective function. It is convex. Had it been concave, the problem would not have been solvable by a linear programming solver. You will see in Chapter 7 (Section 7.2) how to use an integer programming solver to handle this more difficult case.

3.1.1.3 Executable Model

Let's translate this into an executable model. First, assume that the objective function is described by an array D of tuples $(x, f(x))$. This allows us to consider any continuous piecewise linear function. Assume that we are also given a lower bound b for the quantity to produce. This problem is so simple that we know what the solution will be, namely, the lower bound b. See Listing 3-1. But this is meant to illustrate the technique used to solve a piecewise linear function using a linear solver. In the next section, we will use this technique on a more realistic problem.

Listing 3-1. Simplest Example of Piecewise Model (`piecewise.py`)

```
1    def minimize_piecewise_linear_convex(Points,B):
2        s,n = newSolver('Piecewise'),len(Points)
3        x = s.NumVar(Points[0][0],Points[n-1][0],'x')
4        l = [s.NumVar(0.0,1,'l[%i]' % (i,)) for i in range(n)]
5        s.Add(1 == sum(l[i] for i in range(n)))
6        s.Add(x == sum(l[i]*Points[i][0] for i in range(n)))
```

```
7      s.Add(x >= B)
8      Cost = s.Sum(l[i]*Points[i][1] for i in range(n))
9      s.Minimize(Cost)
10     s.Solve()
11     R = [l[i].SolutionValue() for i in range(n)]
12     return  R
```

Line 4 defines our additional variables, one for each breakpoint of the piecewise function. We force that the sum of those be one at line 5. The definition of x at line 6 and its simple bound at 7 will allow us to consider various interesting scenarios.

The objective function is handled in a similar fashion to x at line 8. We solve and return the solution in a table with all the appropriate information to understand what the solver produced.

We will run this code with various bounds to illustrate the types of solution produced. First, Table 3-2 shows a typical run with a solution within a bracket. We set a bound of $x \geq 250$, which is exactly the value obtained.

Note that only two δ are non-zero and that

$$\delta_1 \times B_1 + \delta_2 \times B_2 = 0.37 \times 148 + 0.63 \times 310 = 250,$$

while the cost function is

$$0.37 \times 3552 + 0.63 \times 8088 = 6408.$$

Table 3-2. *Optimal Solution to Convex Piecewise Objective with* $x \geq 250$

Interval	0	1	2	3	4	5	6	Solution
δ_i	0.0	0.3704	0.6296	0.0	0.0	0.0	0.0	$\sum \delta = 1.0$
x_i	0	148	310	501	617	762	959	$x = 250.0$
$f(x_i)$	0	3552	8088	14200	18144	23364	31244	Cost=6408

To illustrate a boundary case, let's set a bound of $x \geq 310$ and the beginning of a bracket, and observe the result in Table 3-3. Notice that only one δ is non-zero in this case and it is set at the maximum weight of one.

Table 3-3. *Optimal Solution to Convex Piecewise Objective with* $x \geq 310$

Interval	0	1	2	3	4	5	6	Solution
δ_i	0.0	0.0	1.0	0.0	0.0	0.0	0.0	$\sum \delta = 1.0$
x_i	0	148	310	501	617	762	959	x=310.0
$f(x_i)$	0	3552	8088	14200	18144	23364	31244	Cost=8088

As a final example of a boundary case, let's force $x \geq 1$. The result is in Table 3-4.

Table 3-4. *Optimal Solution to Convex Piecewise Objective with* $x \geq 1$

Interval	0	1	2	3	4	5	6	Solution
δ_i	0.9932	0.0068	0.0	0.0	0.0	0.0	0.0	$\sum \delta = 1.0$
x_i	0	148	310	501	617	762	959	x=1.0
$f(x_i)$	0	3552	8088	14200	18144	23364	31244	Cost=24

3.1.2 Variations

The first variation is an application of the piecewise approach to non-linear optimization.

3.1.2.1 Non-Linear Function Minimization via Linear Approximations

Since we can solve optimization problems with piecewise linear functions, we can use this approach to approximate convex non-linear functions with piecewise linear functions of increasing accuracy. Here is an example. Say we need to minimize, on the interval [2, 8], the nonlinear function

$$f(x) = \sin(x)e^x$$

We can easily decompose this function into segments on which we interpolate linearly between function values, as in Figure 3-3.

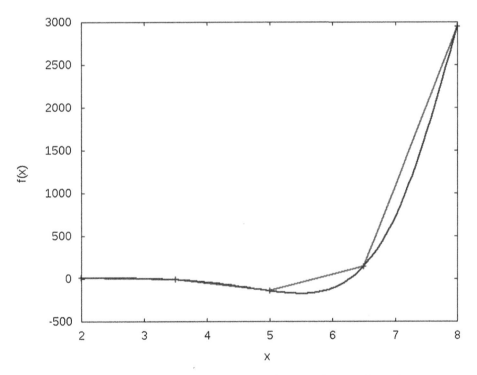

Figure 3-3. *Piecewise approximation of a non-linear function*

We then minimize this piecewise linear approximation. If the solution is accurate enough for our needs, we are done. If not, we zoom in around the solution and approximate the function again using smaller segments. The executable code is shown in Listing 3-2 and is one more instance where using a general-purpose programming language like Python clearly wins over special-purpose modeling languages.

Listing 3-2. Minimizing Non-Linear Functions via Linear Approximations

```
1   def minimize_non_linear(my_function,left,right,precision):
2     n = 5
3     while right-left > precision:
4       dta = (right - left)/(n-1.0)
5       points = [(left+dta*i, my_function(left+dta*i)) for i
        in range(n)]
6       G = minimize_piecewise_linear_convex(points,left)
7       x = sum([G[i]*points[i][0] for i in range(n)])
8       left = points[max(0,[i-1 for i in range(n) \
9                          if G[i]>0][0])][0]
10      right = points[min(n-1,[i+1 for i in range(n-1,0,-1) \
11                         if G[i]>0][0])][0]
12  return   x.SolutionValue()
```

The function `minimize_non_linear` accepts as parameters any Python function, along with an interval of values over which to minimize and a desired precision. At line 4 we compute the length of each sub-interval and we construct a piecewise description of the given function at line 5 which we use as a parameter to our previously described solver (Listing 3-1).

The lines 9 and 11 zoom in on the appropriate sub-interval, which becomes the new interval to be subdivided. The process stops when the interval is smaller than the required precision. In ten very simple lines of codes we leverage the power of a linear solver to minimize non-linear convex functions.

We can see the increasing accuracy of the solution in Table 3-5. Each set of three consecutive rows represent the breakpoints in x, the value of the function at those points, and the interval parameter delta, indicating the optimal bracket. The rightmost two columns are the corresponding optimal x and f(x). We note that x jumps alternatively under and above the final solution, which can be important if one requires an under or over-estimate.

Table 3-5. *Optimal Solution to Non-Linear Minimization*

Interval	0	1	2	3	4	x^*	$f(x^*)$
x_i	2.0	3.5	5.0	6.5	8.0		
$f(x_i)$	6.7	-11.6	-142.3	143.1	2949.2		
δ_i	0.0	0.0	1.0	0.0	0.0	5.0	-142.3
x_i	3.5	4.2	5.0	5.8	6.5		
$f(x_i)$	-11.6	-62.7	-142.3	-159.7	143.1		
δ_i	0.0	0.0	0.0	1.0	0.0	5.8	-159.7
x_i	5.0	5.4	5.8	6.1	6.5		
$f(x_i)$	-142.3	-170.2	-159.7	-72.0	143.1		
δ_i	0.0	1.0	0.0	0.0	0.0	5.4	-170.2
x_i	5.0	5.2	5.4	5.6	5.8		
$f(x_i)$	-142.3	-159.2	-170.2	-171.9	-159.7		
δ_i	0.0	0.0	0.0	1.0	0.0	5.6	-171.9
x_i	5.4	5.5	5.6	5.7	5.8		
$f(x_i)$	-170.2	-172.5	-171.9	-167.8	-159.7		

(*continued*)

Table 3-5. (*continued*)

Interval	0	1	2	3	4	x^*	$f(x^*)$
δ_i	0.0	1.0	0.0	0.0	0.0	5.5	-172.5
x_i	5.4	5.4	5.5	5.5	5.6		
$f(x_i)$	-170.2	-171.7	-172.5	-172.6	-171.9		
δ_i	0.0	0.0	0.0	1.0	0.0	5.5	-172.6
x_i	5.4	5.5	5.5	5.5	5.6		
$f(x_i)$	-171.7	-172.4	-172.6	-172.5	-171.9		
δ_i	0.0	0.0	1.0	0.0	0.0	5.5	-172.6
x_i	5.5	5.5	5.5	5.5	5.5		
$f(x_i)$	-172.4	-172.5	-172.6	-172.6	-172.5		
δ_i	0.0	0.0	1.0	0.0	0.0	5.5	-172.6

3.1.2.2 Non-Convex Piecewise Linear

The most vexing situation occurs when the function to minimize is non-convex. For example, if the unit cost went on decreasing as in Table 3-6 and Figure 3-4 instead of increasing, then the technique presented in this section will fail, as you can see in Table 3-7.

Notice that the sum of the δ is one and the value of the decision variable is correct, but the total cost is nonsensical. It is obtained by a combination of the first and last δ, non-consecutive points. What is happening is that the solver is considering the straight line between $f(0)$ and $f(924)$; this line is below $f(x)$, hence it produces a lower cost value for all intermediate values of x.

Table 3-6. *Example of Non-Convex Piecewise Function*

(From	To]	Unit cost	(Total cost	Total cost]
0	194	18	0	3492
194	376	16	3492	6404
376	524	14	6404	8476
524	678	13	8476	10478
678	820	11	10478	12040
820	924	6	12040	12664

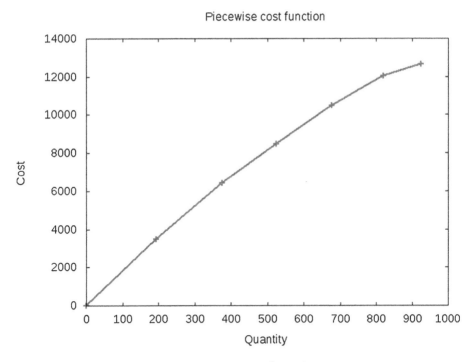

Figure 3-4. *Piecewise non-convex cost function*

Table 3-7. *Incorrect Solution to Non-Convex Objective with $x \geq 250$*

Interval	0	1	2	3	4	5	6	Solution
δ_i	0.7294	0.0	0.0	0.0	0.0	0.0	0.2706	$\sum \delta = 1.0$
x_i	0	194	376	524	678	820	924	x=250.0
$f(x_i)$	0	3492	6404	8476	10478	12040	12664	Cost=3426

As you will see in a later section (Section 7.2 of Chapter 7), all is not lost. The approach taken here can be amended to use an integer solver, adding a few more constraints.

3.2 Curve Fitting

A very common problem is to move from a set of data points to an analytic representation of the same data. Statisticians call this *regression*; applied mathematicians utter *parameter estimation*; and engineers speak of *curve fitting*. I prefer the last expression.[2]

Its most famous and simplest example is illustrated by the following: imagine that we know (or, conjecture, as Galileo was the first to do) that a falling body follows a curve of the form

$$f(t) = a_2 t^2 + a_1 t + a_0,$$

where t represent time, but we do not know the appropriate values for a_0, a_1, and a_2. We run an experiment where we collect the data in Table 3-8.

---

[2]The expression "regression" comes from Francis Galton's original paper about "Re-gression towards mediocrity" and shadows rather than highlights the technique. As for "parameter," what, pray tell, is not a parameter?

Table 3-8. *Example of Data to Fit on a Quadratic*
$f(t) = a_2 t^2 + a_1 t + a_1$

t_i	f_i
0.1584	0.0946
0.8454	0.2689
2.1017	5.8285
3.1966	14.8898
4.056	25.6134
4.9931	38.3952
5.8574	43.5065
7.1474	91.3715
8.1859	119.075
9.0349	115.7737

Since we need to identify the coefficients of our function (the $a_0, a_1, ...$),
we need somehow to minimize a distance from each possible curve to
our data points. Statisticians are fond of using the Euclidean distance or
equivalently, its square,

$$\min \sum_n (\overline{f_i} - f(\overline{t_i}))^2$$

This Least-Squares approach dates from Carl Friedrich Gauss,[3] who
developed it to predict planetary motion. It often makes sense and is very

---

[3]Carl Friedrich Gauss, *Theoria Combinationis Observationum Erroribus Minimis
Obnoxiae (Theory of the Combination of Observations Least Subject to Errors)*
(Philadelphia, PA: Society for Industrial and Applied Mathematics, 1987).

easy to obtain by solving one system of linear equations, the so-called Normal Equations.

Notwithstanding its popularity, the Euclidean distance is not the only valid distance to minimize. Another one is to use the absolute values of the deviations, as in

$$\min \sum_n \left| \overline{f}_i - f\left(\overline{t}_i\right) \right|$$

or even the largest of the absolute values of the deviations, as in

$$\min \ \max_n \left| \overline{f}_i - f\left(\overline{t}_i\right) \right|$$

This latter approach is the most appropriate one when, for instance, we are dealing with tolerances; that is, when all errors must be within some maximal value. We will develop code that can choose, at runtime, between the latter two objective functions.

3.2.1 Constructing a Model

We will describe this rather complex model in stages.

3.2.1.1 Objective Function

Let's assume, with some generality, that we are asked to identify a polynomial of degree k in the variable t. The coefficients $a_0, a_1, \dots, a_k$ are to be determined which minimize either the sum of deviations or the largest deviation between the data points and the polynomial.

The first abstraction here is to think of all of these deviations as some functions, say $e_0, e_2, \dots, e_n$, which we will determine later. In the case of the sum of deviations, the objective is simply

$$\min \sum_i e_i$$

But for the second case, we need an objective that minimizes the maximum deviation:

$$\min_{n} \max \; e_i$$

This latter expression is clearly not a form that fits our framework of linear programs; we can have a min or a max but not both, and we must have one objective function, not a set of them.

The abstract approach to use in cases like this is to move the objectives into the constraints. To illustrate, first we introduce a set of inequalities with a new variable, say e, representing the maximum deviation:

$$e_1 \le e \; \forall i \in [1,n]$$

Second, we state the objective as min e. Since e is an upper bound on all the deviations and we minimize it, we minimize the maximum deviation. Note that, at optimality, at least one of the inequalities will be binding, or else we clearly are not optimal, but most will likely be slack since their deviation will be smaller than the maximum deviation.

3.2.1.2 Constraints

Now we need to express these deviations. We are given a set of couples $(\bar{t}_i, \bar{f})$ representing the measurement at time $\bar{t}_i$ of the putative function f. The deviation for a particular couple is therefore

$$e_i = \left| a_0 + a_1 \bar{t}_i + a_2 \bar{t}_i^2 + \ldots + a_k \bar{t}_i^k - \bar{f}_i \right|$$

That is, the deviation is the absolute value of the difference between the experimental $\bar{f}_i$ and the theoretical displacement $f(\bar{t}_i)$, which is obtained by i evaluating the function at the time $\bar{t}_i$. Why the absolute value? Because the deviation could be positive or negative and we care only about its magnitude.

In terms of the inequalities, we intend to write that we would want a constraint of the form $|f(\bar{t}_i) - \bar{f}| <= e$ but this is not linear. There are at least two different ways to handle this situation. The one to use depends on what information we want to extract from the model's solution.

1. Double the inequalities and bound the deviations.

    Consider the definition of absolute value. $|a| = a$ if a is positive and $-a$ if it is negative. This suggests therefore replacing the inequality $|e_i| <= e$ by two inequalities:

    $$\left|f(t_i) - \overline{f_i}\right| <= e,$$
    $$\left|-f(\overline{t_i}) + \overline{f_i}\right| <= e.$$

    This is a workable approach.

2. Double the variables and find each deviation.

    Note that the "doubling the inequalities" approach will not find the deviation at each point. We simply have a bound of all deviations, a bound which we minimize. What if we want to know each deviation, for instance, to minimize their sum?

One way to find the value of each deviation is to introduce two non-negative variables for each point $(\bar{t}_i, \bar{f}_i)$. Let's call them u_i and v_i and introduce the following equality:

$$f(t_i) - u_i + v_i = \overline{f_i} \tag{3.3}$$

Notice that since the new variables are non-negative, only one of them will be non-zero per equality. That one will equal the deviation (i.e. the difference between the experimental point and the theoretical point).

This "doubling the variables" approach is somewhat more general. If we want to minimize the sum of deviations, we minimize the sum of all u_i and v_i. If we want to minimize the maximum deviation, we add the inequalities

$$u_i \le e,$$
$$v_i \le e,$$

and minimize e.

3.2.1.3 Executable Model

Let's translate this into an executable model seen in Listing 3-3. Assume that we obtain the data in an array of tuples $(\bar{t}_i, \bar{f}_i)$ named D, along with the degree of the polynomial required and an indicator of the distance to minimize (0 for the sum and 1 for the maximum).

Listing 3-3. Polynomial Curve Fitting Model (`curve fit.py`)

```
1   def solve_model(D,deg=1,objective=0):
2     s,n = newSolver('Polynomialufitting'),len(D)
3     b = s.infinity()
4     a = [s.NumVar(-b,b,'a[%i]' % i) for i in range(1+deg)]
5     u = [s.NumVar(0,b,'u[%i]' %  i) for i in range(n)]
6     v = [s.NumVar(0,b,'v[%i]' % i) for i in range(n)]
7     e = s.NumVar(0,b,'e')
8     for i in range(n):
9       s.Add(D[i][1]==u[i]-v[i]+sum(a[j]*D[i][0]**j \
10                         for j in range(1+deg)))
```

```
11      for i in range(n):
12          s.Add(u[i] <= e)
13          s.Add(v[i] <= e)
14      if objective:
15          Cost = e
16      else:
17          Cost = sum(u[i]+v[i] for i in range(n))
18      s.Minimize(Cost)
19      rc = s.Solve()
20      return rc,ObjVal(s),SolVal(a)
```

Line 4 defines the real decision variables, the coefficients of the polynomial. Since we cannot easily set bounds on the coefficients, we use infinity. Lines 5 and 6 define the deviations between the data points and the corresponding theoretical values. This is used at line 8, which corresponds to (3.3). Then we bound the deviations at line 11 by our maximum error variable defined at line 7.

The last element is the choice of objective function. The user can select to minimize the maximum deviation at 15 or the sum of deviations at line 17. These are displayed in Table 3-9 under the headings, respectively, of e_i^{max} and e_i^{sum}.

Table 3-9. *Optimal Solution to Curve Fitting Problem*

t_i	f_i	$f_{sum}(t_i)$	e_i^{sum}	$f_{max}(t_i)$	e_i^{max}
0.1584	0.0946	-0.4382	0.5328	-12.4063	12.5008
0.8454	0.2689	0.3421	0.0731	-8.3251	8.594
2.1017	5.8285	5.8285	0.0	2.0924	3.7362
3.1966	14.8898	14.8898	0.0	14.285	0.6047
4.056	25.6134	24.7951	0.8184	25.8879	0.2744
4.9931	38.3952	38.3952	0.0	40.5766	2.1814
5.8574	43.5065	53.5269	10.0204	56.0073	12.5008
7.1474	91.3715	80.7311	10.6403	82.3995	8.972
8.1859	119.075	106.6547	12.4203	106.5742	12.5008
9.0349	115.7737	130.5102	14.7365	128.2745	12.5008

The data points, along with both solutions (one for minimizing the maximum deviation and one for minimizing the sum of deviation), are displayed in Figure 3-5.

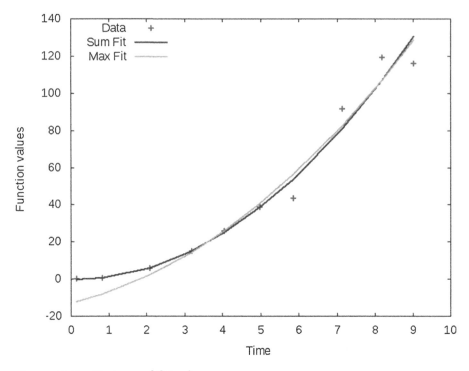

Figure 3-5. *Data and fitted curves*

3.2.2 Variations

The example above is a special case of a technique very useful in practice, the so-called handling of *soft constraints*. Often we would like an equation to be satisfied but we know that it is unlikely to be. Examples abound. Here is one: constructing a model of a scheduling system to produce student schedules at school. In the intelligent manner, we take all the student course choices and their days of availability (I have to work Fridays, so no class then. Or, I work nights; I need day classes only). From all of this data, we wish to construct schedules that work for all students.

Unfortunately, it is unlikely that a schedule accommodating all requests is feasible. The best one can hope for is to satisfy as many students' requests as possible. These become soft constraints and the technique is similar to the one above: we introduce new variables (like the u_i and v_i above) to measure the distance to the ideal (the number of students with unsatisfied requests) and minimize the sum of these.

So, if we aim to satisfy, say

$$a_1x_1 + a_2 x_2 + \dots + a_nx_n = b.$$

But we know that this is unlikely, so we change to

$$a_1x_i + a_ix_2 + \dots + a_nx_n + u - v = b,$$

where u and v are non-negative. Then we add $(u + v)$ to the objective function (assuming this is a minimization problem).

Note that we may need only one of u and v if we already know that the left-hand side will always be either too high or too low with respect to the right-hand side. We need both only when it can deviate in both directions, as in our curve fitting example.

3.3 Pattern Classification Revisited

Recall the classification model from Section 2.5 in Chapter 2: given two sets of data points, benign and malignant cells identified by an expert, we obtained a separating hyperplane that could afterwards be used to mimic the expert by classifying new data into one or the other set. One weakness of our initial approach was that we would not necessarily find the "best" hyperplane by any definition. We stopped as soon as we got one. The result of our first attempt is illustrated in Figure 3-6 where one malignant cell is located exactly on the separating hyperplane. It could just as well have been a benign cell.

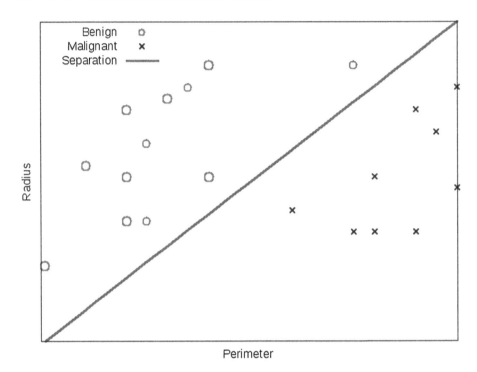

Figure 3-6. *Data and separation maximizing margins*

People in classification therefore tend to prefer a hyperplane that would maximally separate the two sets; that is, one that is equally distant from one set as from the other. This might minimize misclassification later on, assuming that the training set was well chosen.

One way to achieve this maximal separation is to maximize the minimum distance from the training set to the separating hyperplane. This is known as *maximizing the margins* and we now have the tools to perform this maximization. Let's assume that we have run our previous classification model and we know that the two sets are separable by hyperplane $\sum a_j x_j = a_0$. Now we want the best such separation.

How does one compute the distance of a point $\bar{x}$ to hyperplane $\sum a_j x_j = a_0$? By the formula

$$\frac{\left|\sum a_j \overline{x}_j - a_0\right|}{\sqrt{\sum a_j^2}}$$

Since we intend to maximize the minimum of this value over all data points yet have no need for the actual value, the denominator is irrelevant and we might as well simply consider the numerator, a fortuitous condition as we cannot yet handle general nonlinear functions. This numerator is an absolute value. We will therefore use the double positive variable trick and introduce for each data point $\bar{x}$ a triplet of constraints

$$\sum a_j \overline{x}_j - u + l = a_0$$
$$e <= u$$
$$e <= l$$

where the first constraints will force either u or l to measure the value of the numerator in the distance formula. The two inequalities will have the variable e lower bound both. We then only have to maximize the value of e to achieve our goal.

3.3.1 Executable Model

The translation into an executable model is shown in Listing 3-4. Lines 4 to 7 define the new positive variables to hold the distance from each data point to the separating hyperplane. Line 8 is the same as before, the coefficient of the hyperplane we are searching for. The rest of the model is identical to our previous classification model with the addition of three lines at 12 and at 17 fixing the distance constraint and establishing the lower bound e on them. The objective function is now to force this lower bound upward, maximizing the minimum distance.

Listing 3-4. Maximizing the Margins (`margins.py`)

```
1    def solve_margins_classification(A,B):
2      n,ma,mb=len(A[0]),len(A),len(B)
3      s = newSolver('Classification')
4      ua = [s.NumVar(0,99,") for _ in range(ma)]
5      la = [s.NumVar(0,99,") for _ in range(ma)]
6      ub = [s.NumVar(0,99,") for _ in range(mb)]
7      lb = [s.NumVar(0,99,") for _ in range(mb)]
8      a = [s.NumVar(-99,99,") for _ in range(n+1)]
9      e = s.NumVar(-99,99,")
10     for i in range(ma):
11       s.Add(0 >= a[n]+1-s.Sum(a[j]*A[i][j] for j in
           range(n)))
12       s.Add(a[n]==s.Sum(a[j]*A[i][j]-ua[i]+la[i]for j in
           range(n)))
13       s.Add(e <= ua[i])
14       s.Add(e <= la[i])
15     for i in range(mb):
16       s.Add(0 >= s.Sum(a[j]*B[i][j] for j in range(n))-a[n]+1 )
17       s.Add(a[n]==s.Sum(a[j]*B[i][j]-ub[i]+lb[i]for j in
           range(n)))
18       s.Add(e <= ub[i])
19       s.Add(e <= lb[i])
20     s.Maximize(e)
21     rc  = s.Solve()
22     return  rc,SolVal(a)
```

The result, on the same data set as in Section 2.5, shows the new and improved separating hyperplane, equidistant from the closest points both in the malignant set and the benign set, which is as good a separation as we can hope for. But beware: this is only as good as the training set. If this set was biased in any way, the separation will be just as biased.

CHAPTER 4

Linear Network Models

Six degrees of separation!

This meme and the play, movies, and games it generated did more to introduce elements of network theory to the general public than years of public education. Movie buffs play the Kevin Bacon Game, trying to link two actors through movies in which they appeared with Kevin Bacon. Mathematicians proudly announce their *Erdös number* (when they have one), the number of co-written papers away they stand from a paper co-written with the famous Paul Erdös.[1] Throughout this chapter networks play an essential role in visualizing a problem.

A network is an object composed of nodes (people, in our examples) and arcs (indicating the presence of a relationship). It is a tool mathematicians invented hundreds of years ago to help model situations[2] and solve problems. We will use networks to help us construct optimization models. In a sense, we are doing meta-modeling.

Network-based optimization models often share, along with the structural description, an interesting characteristic: if the input data are all integers, then there is an integral optimal solution. Moreover, the solver

---

[1]`www.oakland.edu/enp/`

[2]The first instance of such a model is usually attributed to Euler.

© Serge Kruk 2018
S. Kruk, *Practical Python AI Projects*, https://doi.org/10.1007/978-1-4842-3423-5_4

will find it. This is a useful property as it allows for modeling of countable items (people, trucks, data packets) as well as measurable items (money, time, water).

It is important to realize the importance of integrality in modeling. Let's imagine a complex problem involving the flight of a space shuttle. The constraints involve weight, amount of fuel, amount of oxygen, work to do while in orbit, etc. There are thousands, maybe hundreds of thousands, of variables and constraints. If such a problem asks "How many astronauts can be carried in the shuttle?" it is unlikely that NASA (or SpaceX) would accept "Two and a half astronauts" as the optimal answer.

Crucially, we must dispense with the tempting but wrong work-around: rounding. **It is rarely the case that rounding helps.** If we round a fractional solution, many (maybe even all) constraints might be violated. Round up the solution of our space shuttle instance and the weight constraint might prevent lift-off; round down and the astronauts may fail to accomplish all the required tasks. To be fair, there are problems where rounding is acceptable, but those are either boring, or the solution is obvious, or both.

4.1 Maximum Flow

Network-related problems often have a structure where the integrality of the solution is guaranteed "for free." We can do nothing except recognize that the problem falls in that special category and be merry. The goal of this section is to recognize and model problems with this special structure.

The prototypical, overt example is the *network maximum flow* (*maxflow*) problem where some substance *flows* from some source(s) to some destinations(s) on capacitated channels and we try to maximize the amount of flowing.

The substance *flowing* does not have to be material, as water, oil, or even electricity. It could be data packets flowing through a network of fiber optics cables. Imagine, for instance, that you are trying to establish how many concurrent video streams you can send from your servers to your viewers. This fits nicely in the context of a maximum flow problem.

To consider the simplest problem abstractly, let's assume a network as described by Figure 4-1 where each arc has the noted capacity and we are trying to send as much as possible from the nodes marked as sources (-*S*) to the nodes marked as sinks (-*T*).

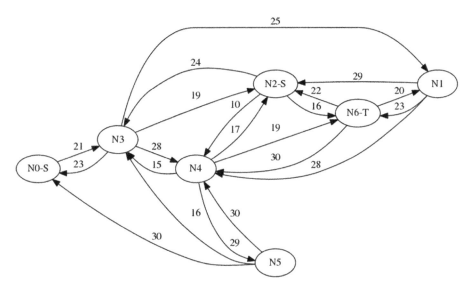

Figure 4-1. *Visual representation of a network flow problem instance*

4.1.1 Constructing a Model

What we need to decide in this problem is the amount to deliver from the source to the sink and through which arcs of the network.

4.1.2 Decision Variables

The simplest, most natural way to answer the question is to introduce a two-dimensional variable. The first dimension indicates the origin node and the second dimension indicates the destination node; the value of the variable will be the amount of substance flowing on the arc between the two, or

$$x_{i,j} \quad \forall i \in N, \forall j \in N$$

For example, if $x_{2,3} = 35$, it will mean that we should send 35 units from node 2 to node 3.

4.1.2.1 Objective

The objective is to maximize the amount flowing from the sources to the sinks. So the sum of either what is coming out of the sources (say set S) or the sum of what goes into the sinks (say set T) will work. This suggests either of these objective functions

$$\max \sum_{i \in s} \sum_{j \in N} x_{i,j} \quad \max \sum_{j \in T} \sum_{i \in N} x_{i,j}$$
or

But, since we allow multiple sources and nothing prevents a source from sending to another source for further transport, we should be careful to maximize the "net" flow out of the sources, or

$$\max \sum_{i \in S} \left(\sum_{j \in N} x_{i,j} - \sum_{j \in N} x_{j,i} \right) \tag{4.1}$$

The corresponding net flow into the sinks should be obvious.[3]

This objective function (4.1) is enough to get a working model, but we have two minor annoyances to consider: chained sources and cycles.

The first problematic case is illustrated by Figure 4-2. It is clear that since the capacity of the arc going into the sink is 2, exactly two units will flow. But these two units could come from source *N1-S* or one could come from *N0-S* and a second from *N1-S*.

Figure 4-2. *Problematic chaining of sources*

The second problematic case occurs when there is a cycle involving a source. Then, given any flow *f* around that cycle, if that flow is not as large as the capacity on that cycle, then there is another flow, *f* + 1, with exactly the same objective value. Consider Figure 4-3 as an example. An optimal solution could send 10 units from the source to the sink through the intermediate node or up to 20 units from the source, with up to 10 units flowing back from the intermediate node to the source.

Figure 4-3. *Problematic cycles*

These two cases illustrate multiple optimal solutions with, of course, exactly the same optimal value. There is little chance that, for any application, this multitude of optimal flows represents a feature. It will likely be considered a nuisance, especially if two different solvers or two

---

[3]Note that there may be applications where the flow out of the source is to be maximized without regards to the flow in.

different runs of the same solver return two different flows! How can we ensure that the solver consistently returns the same flow?

We could decide that, among all optimal flows, we want the flow with as little flow into the sources as possible. This is a *dual objective*: maximize the net flow out of the sources, and minimize the flow into the sources. This idea of a dual objective, or more generally of multiple objectives, often occurs in practice and often for the same reason as here: the need to determine, among multiple optimal solutions according to one criterion, the most preferable one according to a secondary criterion.

Since we need to maximize one object and minimize another, all at one go, we need another trick, the reversal:

$$\max f(x) \Leftrightarrow \min - f(x)$$

We can always replace a minimization problem by a maximization problem, and vice versa. And now we can add the two objectives: maximizing the net flow as in equation (4.1) and minimizing the inflow, or maximizing $-\sum_{i \in S} \sum_{j \in N} x_{j,i}$. After simplification, it looks like

$$\max \sum_{i \in S} \left(\sum_{j \in N} x_{i,j} - 2^* \sum_{j \in N} x_{j,i} \right) \tag{4.2}$$

In the case of our chaining example, we would maximize $x_{0,1} + x_{1,2} - 2^* x_{0,1}$, or $x_{1,2} - x_{0,1}$ which would force a solution where all the flow is issued from N1-S. In our cyclic example, this would yield $x_{0,1} + x_{1,2} - x_{1,0}$ which guarantees that no flow comes back into the source and it emits 10 units.

4.1.2.2 Constraints

The only constraint type is known as *conservation of flow*: for every node that is neither a source nor a sink, whatever flow goes in must come out, or

$$\sum_{j \in N} x_{i,j} = \sum_{j \in N} x_{j,i} \forall i \in N \backslash \{S \cup T\} \tag{4.3}$$

Since the objective function will force flow out of the sources or, equivalently, into the sinks, the conservation of flow will take care to move the material from sources to sinks.

4.1.2.3 Executable Model

Let's translate this into an executable model. To make the model general enough to solve all problems of this type, we will assume that the input is a two-dimensional array called *C*, indexed by nodes, containing the capacity of the arc between two nodes. We will also assume arrays, one of sources *S* and one of sinks *T*.

To allow some flexibility in the choice of objective function and illustrate the occurence of multiple optimal solutions, we add a final parameter, unique. If this parameter is set to True, the model will run with objective function (4.2), which will maximize the net flow while minimizing the flow into the sources. If set to False, it will simply maximize the flow out of the sources. See Listing 4-1.

Listing 4-1. Maximum Flow Model (maxflow.py)

```
1   def solve_model(C,S,T,unique=True):
2       s,n = newSolver('Maximumuflowuproblem'),len(C)
3       x=[[s.NumVar(0,C[i][j],")for j in range(n)]
         for i in range(n)]
4       B=sum(C[i][j] for i in range(n) for j in range(n))
5       Flowout,Flowin   =   s.NumVar(0,B,"),s.NumVar(0,B,")
```

```
6     for i in range(n):
7       if i not in S and i not in T:
8         s.Add(sum(x[i][j] for j in range(n)) == \
9         sum(x[j][i] for j in range(n)))
10     s.Add(Flowout == s.Sum(x[i][j] for i in S for j
       in range(n)))
11     s.Add(Flowin == s.Sum(x[j][i] for i in S for j
       in range(n)))
12     s.Maximize(Flowout-2*Flowin if unique else
       Flowout-Flowin)
13     rc = s.Solve()
14     return rc,SolVal(Flowout),SolVal(Flowin),SolVal(x)
```

Line 3 defines the two-dimensional variable where the first index specifies the origin and the second, the destination. Line 9 ensures that we conserve flow across nodes that are neither sources nor sinks. The objective function at line 12 computes the total flow and indicates that we should maximize that quantity.

We return the total flow out of the sources and the total flow into the sources. The output of the model is displayed in Table 4-1 and Table 4-2 where each corresponds to the choice unique at, respectively, False and True.

Table 4-1. *Optimal Solution Maximizing Net Flow*

71-13	N0-S	N1	N2-S	N3	N4	N5	N6-T
N0-S				21.0			
N1							23.0
N2-S				24.0	10.0		16.0
N3		23.0	13.0		9.0		
N4							19.0
N5							
N6-T							

Table 4-2. *Optimal Solution Maximizing
Net Flow and Minimizing Inflow*

58-0	N0-S	N1	N2-S	N3	N4	N5	N6-T
N0-S				8.0			
N1							23.0
N2-S				24.0	10.0		16.0
N3		23.0			9.0		
N4							19.0
N5							
N6-T							

Here is the interesting phenomenon: all solutions are integers, yet we did not impose an integrality constraint. This is a consequence of two elements. First, the structure of the problem, which guarantees that if there is an optimal solution, there is an integral solution.[4] Second, the solution technique of all solvers, which will either only consider integer solutions (simplex solvers) or will move from a fractional solution to an integer one (interior-point solvers) before returning to the caller. The reader is encouraged to tweak the numbers to verify that, if the problem is feasible, the solver will find an integral solution.

4.1.3 Variations

One useful application is to model assignment problems. These come in multiple flavors of which here is one: imagine that we have a certain number of workers (they could be people, machines, or cores in a desktop computer) and a certain number of jobs to accomplish (papers to push, widgets to build, or programs to execute). We construct a network with one source connected by an arc of unit capacity directed to each worker. These workers are connected by an arc to job nodes, the sinks, but only they are capable of executing the given job. Maximizing the flow will assign workers to jobs optimally in the sense that the most jobs will get done. We read the assignments by looking at the arcs with non-zero flow between workers and jobs. See Figure 4-4.

---

[4]The theoretically-minded reader will research "total unimodularity."

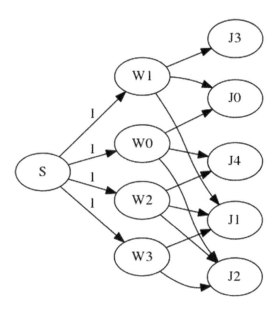

Figure 4-4. *Assigning workers to jobs*

4.2 Minimum Cost Flow

There is a second class of problems where the integrality of the solution is guaranteed "for free:" the *minimum cost flow* problems (*mincost*).

Here is the prototypical example. Solar-1138 Inc. has a set of clean power plants supplying the needs of multiple cities. Each power plant has a maximum capacity, so it can supply a limited number of kilowatt-hours (kW-h). Each city has a peak demand and all cities peak at roughly the same time. Therefore, the sum of peak demands is the quantity that the power plants need to accommodate. The cost of sending one kW-h from a plant to a city varies according to the plant, the city, the delivery infrastructure, and distance between plant and city. This cost has been arrived at by considering production of the power and maintenance of the plant and the power lines.

Table 4-3 has the cost of delivery between plant and city, whenever possible, in dollars per kW-h. The maximum supply of each plant and the peak demand of each city are in kW-h.

Table 4-3. *Example of Electrical Distribution Cost*

From/To	City 0	City 1	City 2	City 3	City 4	City 5	City 6	Supply
Plant 0	23		19	25	14		22	551
Plant 1	16			20	23	13	23	689
Plant 2	22	18	11		20	13	24	634
Demand	288	234	236	231	247	262	281	

The question to answer is "How much power should be sent from each plant to each city, satisfying peak demand while minimizing cost?"

4.2.1 Constructing a Model

What we need to decide in this problem is the amount of power to deliver from each plant to each customer. It may be that a plant sends power to none, some, or all cities, and we need a way to indicate this. As a visual aid, consider the bipartite[5] graph shown in Figure 4-5 where plants are the top nodes, cities the bottom, and the arcs are power lines annotated by the cost of carrying power along that particular transmission line.

---

[5]Bipartite means that there will never be any arcs between the top nodes or between the bottom nodes. You will see in the next section a more general problem.

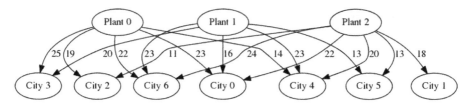

Figure 4-5. *Caption needed*

4.2.1.1 Decision Variables

With this image in mind, a solution to the problem would be, for each plant-city pair, the amount of power delivered by plant to city. The simplest, most natural way to do this is to introduce a two-dimensional variable. The first dimension indicates the origin (which plant of the set of plants P) and the second dimension indicates the destination (which city out of the set of cities C), or

$$x_{i,j} \quad \forall i \in P, \forall j \in C$$

For example, if $x_{2,3} = 35$, it means we should send 35 kW-h from plant 2 to city 3.

These multi-dimensional variables, and you will see many of them, are simply a short-hand notation for one variable per each combination plant-city. So if we have 3 plants and 4 cities, we are really introducing 3.4 = 12 decision variables. This is somewhat wasteful of memory as there are not always paths between each plant and each city, but you will see how to avoid the waste if it becomes an issue.

4.2.1.2 Objective

The objective is to minimize cost in dollars of delivery. For this we will need a parameter for the cost. Let's assume $C_{i,j}$ indexed exactly as the decision variable. The objective will therefore be

$$\min \sum_i \sum_j C_{i,j} x_{i,j} \tag{4.4}$$

4.2.1.3 Constraints

The constraints are of two types, closely related: supply and demand. To refer to them, we introduce the following parameters, S_i, $i \in P$ and D_j, $j \in C$, indicating the supply possible from plant i and the demand required by city j.

Each plant has a maximum production capacity. We need to respect this maximum. So for each plant, we must cap the sum of the power delivered from that plant by an availability constraint, as in

$$\sum_j x_{i,j} \leq S_i \quad \forall i \in P$$

Notice the inequality: we are not forcing the amount delivered to be at capacity, but only to be at most the supply capacity.

The city demands are similar except that they must be met. Therefore,

$$\sum_i x_{i,j} = D_j \quad \forall j \in C$$

Here we have an equality constraint. If we make a modeling error and put an inequality, say $\leq$, the optimal solution will be all zero. On the other hand, we could have a $\geq$. In this case, it would change nothing to the solution since we are minimizing the total cost.

4.2.1.4 Executable Model

Let's translate this into an executable model. To make the model general enough to solve all problems of this type, we will assume that the costs, demands, and supply capacities are given in a two-dimensional array called *D* structured as in Table 4-3, where a zero cost indicates that there are no power lines between that plant-city combination.

In more general terms than "plants" and "cities," we can view each row as representing a producer and each column a consumer, except that the last row represents the demand and the last column the supply. The "product" exchanged between producers and consumers can be anything, not only divisible quantities like kW-h or liters of water, but trucks, flowers, data packets, or people. The optimal solution will never contain fractions of people. See Listing 4-2.

Listing 4-2. Power Distribution Model (`mincost.py`)

```
1    def solve_model(D):
2        s = newSolver('Mincostuflowuproblem')
3        m,n = len(D)-1,len(D[0])-1
4        B = sum([D[-1][j] for j in range(n)])
5        G = [[s.NumVar(0,B if D[i][j] else 0,") for j in
             range(n)] \
6             for i in range(m)]
7        for i in range(m):
8          s.Add(D[i][-1] >= sum(G[i][j] for j in range(n)))
9        for j in range(n):
10         s.Add(D[-1][j] == sum(G[i][j] for i in range(m)))
11       Cost=s.Sum(G[i][j]*D[i][j] for i in range(m)for j in
             range(n))
12       s.Minimize(Cost)
13       rc  = s.Solve()
14       return rc,ObjVal(s),SolVal(G)
```

Line 6 defines the two-dimensional variable where the first index specifies the producer and the second, the consumer. Since we know that if a particular producer-consumer pair has no channel between, the entry is zero, so we use this to collapse the range of the variable to zero. A good solver will use this information to eliminate these variables before doing any other work.

Line 8 ensures that we supply no more than each plant can produce while line 10 ensures that the peak demands are satisfied. The objective function at line 11 computes the total cost and indicates that we should minimize that quantity.

The output of the model is displayed in Table 4-4. The reader can verify that the total column is below or at the maximum that each plant can produce, while the total row is exactly the required peak demand of each consumer.

Table 4-4. *Optimal Solution to the Power Distribution Problem*

From/To	City 0	City 1	City 2	City 3	City 4	City 5	City 6	Total
Plant 0					247		281	528
Plant 1	288			231		170		689
Plant 2		234	236			92		562
Total	288	234	236	231	247	262	281	

Here is again the interesting phenomenon: all solutions are integers, yet we did not impose any integrality constraints.

4.2.2 Variations

The simplest variation is to have capacities on the arcs. Then we need to insure that no flow goes above the capacity. Assuming we have the capacities in matrix A, this is simply a question of adding a constraint of the form

$$x_{i,j} \leq A_{i,j} \ \forall i \in P, \forall j \in C$$

An interesting variation involves spreading the sources. To minimize risks, for instance, we might not want to satisfy more than some fraction of the demand from one source. Say we decide that no city's demand may be satisfied at more than 60 percent from a single source, we could add a constraint of the form

$$x_{i,j} \leq 0.6D_j \ \forall i \in P, \ \forall j \in C$$

The reader is encouraged to add this constraint and note that the optimal value will not be as low as it is without the constraint. Moreover, the solution might not be integral anymore. This simple additional constraint destroys the property that guarantees integrality. We must declare the decision variable integral (with the consequent increase in complexity and solution time) to guarantee an integral solution.

Instead of material that flows, the problem sometimes appears as an *assignment* question: given a set of workers with specific skills and hourly wages and a set of jobs, which worker do you assign to which job to minimize cost?

For example, consider a consulting firm with three teams based in different cities as well as three customers at different sites. Since the cost of travel varies from team to customer site, we want to minimize total travel cost. In this case, the demand and supply are simply one since we want one team per customer site and one customer assigned to each team. See Table 4-5.

Table 4-5. *Caption Needed*

	Customer 0	**Customer 1**	**Customer 2**	**Supply**
Team 0	25	30	20	1
Team 1	20	15	35	1
Team 2	18	19	28	1
Demand	1	1	1	

4.3 Transshipment

A more general type of problem that can be modeled as a network flow problem is the *transshipment* problem. The characteristics of such a problem are a set of nodes with a cost of transporting between each pair; a subset of the nodes is suppliers and another subset is consumers. The remaining nodes can be used to carry the material but neither produces nor consumes, hence the moniker *transshipment*.

Table 4-6 has, for example, the cost of delivery between each pair of nodes. A blank indicates that there is no path between two nodes. The last column indicates the amount that a node can produce, if any; the last row is the demand of each node, if any. Note that the sum of demands should, in general, be the same as the sum of supplies or else the problem is infeasible.

Table 4-6. *Example of Transshipment Distribution Cost Over a Network*

From/To	N0	N1	N2	N3	N4	N5	N6	N7	Supply
N0					17	10	19		
N1	23			28		23			
N2	29				30	25	25		680
N3					17	15	19	29	
N4		16							
N5	22				25			18	540
N6	25	29	16			22			
N7			30		10		27		
Demand	241			164	239		152	424	

Transshipment problems are often depicted visually as in Figure 4-6, corresponding to the data in Table 4-6, where the arrows are annotated with the cost of transportation of the material and where the nodes contain a positive number indicating supply value and/or a negative number indicating a demand value.

Note that this is clearly a generalization of the mininimum cost flow problem as there could be arcs between any two nodes, which means, for instance, that a source node could receive whatever product is flowing through the network, add it to its production, and send the result off towards another node, whether a consumer or a transshipment node.

4.3.1 Constructing a Model

What we need to decide in this problem is the amount of material to deliver from each node with a positive supply to each node with a positive demand. The simplest, most natural way to model this is to introduce a two-dimensional variable. The first dimension indicates the origin and the second dimension indicates the destination. The variable itself will contain the amount to ship. We will assume that N is the set of nodes to get

$$x_{i,j} \quad \forall i \in N, \forall j \in N$$

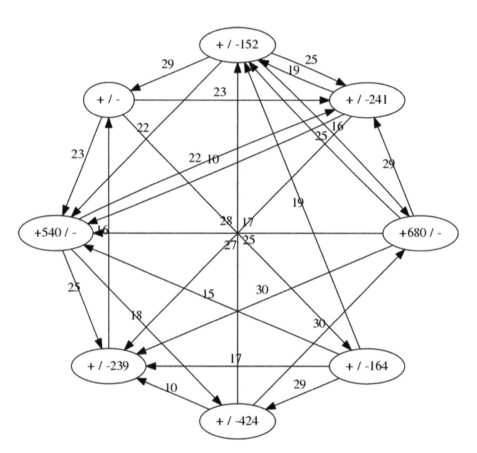

Figure 4-6. *Transshipment example viewed graphically*

For example, if $x_{2,3} = 35$, it means we should send 35 units from node 2 to node 3.

4.3.1.1 Objective

The objective is to minimize cost in dollars of delivery. For this we will need a parameter for the cost. Let's assume $C_{i,j}$ indexed exactly as the decision variable. The objective will therefore be

$$\min \sum_i \sum_j C_{i,j} x_{i,j}$$

4.3.1.2 Constraints

In the previous minimum cost flow problem, we had two types of constraints: one for the producing nodes which stated that the outflow was equal to the supply value, and another one for the consuming nodes, stating that the inflow was equal to the demand value. We could use these constraints here, in addition to a third constraint for intermediate nodes, those without any demand or supply value, stating that the inflow must equal flow.

While it is possible to treat each of type of nodes (producers, consumers, and intermediate) differently, it is simpler to notice that one constraint, appropriately general, will hold each node. Namely, that the flow into the node (f_{in}) minus the flow out of the node (f_{out}) must equal the demand (D) minus the supply (S), or

$$f_{in} - f_{out} = D - S.$$

Note that in the case of pure sources and sinks, the equation reduces to the constraints we used in the mininimum cost flow model:

$$-f_{out} = -S \text{ Special case for sources}$$

$$f_{in} = D \text{ Special case for sinks}$$

Let's assume that S_i and D_i are, respectively, the supply and the demand of node i, keeping in mind that for producing nodes only supply is non-zero; for consuming nodes, only demand is non-zero; and for intermediate nodes, both are zero. We get

$$\sum_j x_{j,i} - \sum_j x_{i,j} = D_i - S_i \ \forall i \in N \tag{4.5}$$

This constraint is known as a general *conservation of flow* constraint. It is the only constraint we need, but it must hold at every node.

4.3.1.3 Executable Model

Let's translate this into an executable model. To make the model general enough to solve all problems of this type, we will assume that the costs, demands, and supply capacities are given in a two-dimensional array called D structured as in Table 4-6. Each entry {i,j} represents the cost of transporting from node i to node j, except that the last row represents the demand and the last column the supply. See Listing 4-3.

Listing 4-3. Transshipment Distribution Model (transship dist.py)

```
1    def solve_model(D):
2        s = newSolver('Transshipmentuproblem')
3        n = len(D[0])-1
4        B = sum([D[-1][j] for j in range(n)])
5        G = [[s.NumVar(0,B  if D[i][j] else  0,") \
6            for j in range(n)] for i in range(n)]
```

```
7      for i in range(n):
8        s.Add(D[i][-1] - D[-1][i] == \
9        sum(G[i][j] for j in range(n))-sum(G[j][i]for j in
         range(n)))
10       Cost=s.Sum(G[i][j]*D[i][j] for i in range(n)for j in
         range(n))
11       s.Minimize(Cost)
12       rc = s.Solve()
13       return rc,ObjVal(s),SolVal(G)
```

Line 6 defines the two-dimensional variable where the first index specifies the producer and the second, the consumer. The range of a variable is from zero to either the total demand or to zero to ensure that we don't use a route that does not exist. In data D, the absence of a cost at entry i,j indicates that there is no direct route between i and j.

The generalized conservation of flow constraint corresponding to (6.5) is implemented on line 7. The objective function at line 10 computes the total cost and line 11 indicates that we should minimize that quantity.

The output of the model is displayed in Table 4-7. Note that the solution is again entirely integral, even though we did not enforce integrality.

The reader can verify that the difference between the entry in the total column of a given node less the entry in the total row of the same node is equal to the difference of the supply and the demand of that node. This is especially interesting for nodes that receive more than their demands and reroute whatever they do not use. Even for very small problems, those are not solutions can be easily guessed.

Table 4-7. *Optimal Solution to the Transshipment Problem*

From/To	N0	N1	N2	N3	N4	N5	N6	N7	Out
N0									
N1				164					164
N2	125				403		152		680
N3									
N4		164							164
N5	116							424	540
N6									
N7									
In	241	164		164	403		152	424	

4.3.2 Variations

- One possible variation is when the supply and demand
  are not balanced. It could be that the producing nodes
  have a maximum production capacity that they need
  not meet; only the demands must be satisfied. In
  this case, these nodes must be treated separately
  and, instead of the generalized conservation of flow
  constraint, we indicate that the flow out less the flow in
  must be at most the supply, so

$$\sum_j x_{i,j} - \sum_j x_{j,i} \leq S_i \ \forall \{i \in N \mid S_i > 0\}$$

  Nothing else needs to change because we are
  minimizing cost and satisfying demands, so we will
  get the optimal solution.

- The reverse situation is also possible, though unlikely, in which case we must treat the demand nodes separately and ensure that the flow in less the flow out must be at most the demand value.

- Another simple variation is to have capacities on the arcs, limiting the amount flowing through them. In this case, the additional constraints are, assuming a matrix of capacity C,

$$x_{i,j} \leq C_{i,j}$$

4.4 Shortest Paths

Now let's consider the problem Google faces every time someone asks Google Maps to find a path from point A to point B: the shortest path problem (either shortest according to distance or according to time). It may surprise the reader that this too can be modeled and solved very efficiently as a network flow problem.

Here is the abstracted situation: we are given a two-dimensional array of distances between a set of points as exemplified by Table 4-8. This is called the *distance matrix*. It could be distances in thousands of kilometers between cities if we are considering a planetary scale problem or times in minutes between city street intersections if we are looking for a bike path taking into consideration the path elevation. In addition to the array of distances, we could be given a start and an end point but, in their absence, we will assume that the array has been ordered so that we need a path from the first point to the last.

The task is to find a sequence of points between the start and the end that minimizes the corresponding sum of entries in the array. This is called a shortest path, no matter what the units are. Note that we do not say *the* shortest path as there may be many paths with the same shortest total distance. So if we go through the sequence 0,3,2,5, for instance, our total distance will be the sum of $M_{0,3} + M_{3,2} + M_{2,5}$.

Table 4-8. *Example of a Distance Matrix*

	P0	P1	P2	P3	P4	P5	P6	P7	P8	P9	P10	P11	P12
P0		46	17	24	51								
P1	46				31	33		54					
P2		38			34	31			51				
P3	24				33			17	49	31			
P4	51					4			18	39	60		
P5	48				4		4	27		35	57	51	
P6				33	1						59		
P7		54	26		32	27	31			14	42	66	
P8			51	49	18	20	17	43			57		32
P9					39	35		14			28		
P10					60							58	6
P11									32	61	58		56
P12									59		56		

4.4.1 Constructing a Model

What we need to decide in this problem is the sequence of points to choose to go from start to end. This is a subset of the given points (say P) along with an order in which to traverse them. It turns out that the most efficient approach is to picture a graph with the points as nodes and with the distances as weights on the arcs. Choosing a path on the graph will correspond to a path on the original map.

At first glance, this may not seem natural. It may not even be clear how to construct a decision variable to hold the subset of points and the order in which to visit them. Here is the trick: for each arc on the graph we have a decision variable that will take on exactly one of two values, either zero if we do not take this arc or one if we do. Therefore,

$$x_{i,j} \in [0,1] \; \forall i \in P, \; \forall j \in P$$

To pursue our example path of 0,3,2,5, we will have decision variables $x_{0,3}$, $x_{3,2}$, and $x_{2,5}$ at value one and all other arc variables at value zero. The reader should see the parallel with the other flow problems by thinking of a network where all arcs have a capacity of one: an integral solution will be a flow of value of one on some sequences of adjacent arcs and zero on all others.

The objective function is correspondingly simple. Assuming that the distance matrix is D,

$$\min \sum_i \sum_j D_{i,j} x_{i,j},$$

how do we ensure that we choose a sequence of adjacent arcs from the start point to the end point? By modeling this as a unit flow through the graph where the start point is a source of value one and the end point is a sink of value one. All we need is the usual flow conservation constraints we used previously.

The executable model is seen in Listing 4-4, where we assume a distance matrix D with optional starting and ending points. We could use our existing code for flow problems but, in this case, as a courtesy to the caller, we will write a special-purpose code to help the call and to return a meaningful answer. After all, we as modelers are thinking of this problem as a flow on a graph, but the caller is thinking of a shortest path! Let's not burden him with our unnatural decision variables. Not to mention that there may be a million points, hence a trillion[6] decision variables and yet the solution, from the caller's perspective, may be only a minuscule fraction of those variables.

Listing 4-4. Shortest Path Model (shortest path.py)

```
1    def solve_model(D,Start=None, End=None):
2      s,n = newSolver('Shortestupathuproblem'),len(D)
3      if Start is None:
4        Start,End = 0,len(D)-1
5      G = [[s.NumVar(0,1  if D[i][j] else  0,") \
6            for j in range(n)] for i in range(n)]
7      for i in range(n):
8        if i == Start:
9          s.Add(1 == sum(G[Start][j] for j in range(n)))
10         s.Add(0 == sum(G[j][Start] for j in range(n)))
11       elif i == End:
12         s.Add(1 == sum(G[j][End] for j in range(n)))
13         s.Add(0 == sum(G[End][j] for j in range(n)))
14       else:
15         s.Add(sum(G[i][j] for j in range(n)) ==
16             sum(G[j][i] for j in range(n)))
17       s.Minimize(s.Sum(G[i][j]*(0 if D[i][j] is None else D[i][j]) \
18                 for i in range(n) for j in range(n)))
19     rc  = s.Solve()
20     Path,Cost,Cumul,node=[Start],[0],[0],Start
21     while rc == 0 and node != End and len(Path)<n:
22       next = [i for i in range(n) if SolVal(G[node][i]) == 1][0]
23       Path.append(next)
```

---

[6]If the reader reads American; a billion if English.

```
24          Cost.append(D[node][next])
25          Cumul.append(Cumul[-1]+Cost[-1])
26          node = next
27      return rc,ObjVal(s),Path,Cost,Cumul
```

On line 3 we set the start and end nodes to be the first and last if the caller did not specify any. Line 6 defines the decision variable. We apply a little trick to the range: we know that if the distance matrix has a zero entry, it means that there is no path between two points. In that case, we give a range of [0, 0], which forces this variable to be zero. In the other cases, the range will be [0, 1]. Notice that this range allows fractions but, again because of the structure of the constraints in a flow problem, no variable will ever have a fractional value. They will all be either 0 or 1.

At lines 9 and 12 we set the supply to be one at the start node and the demand to be one at the end node. At all other nodes (line 16) conservation of flow ensures that whatever goes in comes out. This will produce a solution consisting of a continuous path from start to end.

The objective function at line 18 has the same structure as all the flow problem examples: the product of the cost (here a distance) with the indicator variable of the arc used.

After we solve the problem, we process the solution to return to the caller something smaller, and potentially more meaningful, than our decision variable: a sequence of jumps, from point to point, along with the distance of each jump. It is the job of the modeler to hide the tricks required to solve a problem and provide meaningful solutions to the caller. A solution corresponding to the example above is shown in Table 4-9.

Table 4-9. *Optimal Solution to Shortest Path Problem*

Points	0	3	7	9	10	12
Distance	0	24	17	14	28	6
Cumulative	0	24	41	55	83	89

4.4.2 Alternate Algorithms

If the reader is aware of Dijkstra's algorithm, he may wonder why we create a linear programming model for shortest paths, especially since a fast implementation of Dijstra's algorithm might be faster. The answer is that in the real life of a modeler we rarely have to solve pure shortest paths (or pure anything, really). For the vast majority of situations outside of textbooks, the kernel of the problem might be a shortest path, but there are bound to be multiple additional considerations. And adding these considerations in the form of additional constraints to a basic shortest path linear program is often a simple matter. In contrast, trying to modify an implementation of Dijkstra (assuming that we even have access to the source code) might prove considerably more difficult, if at all possible.

4.4.3 Variations

- It may be that, instead of minimizing the sum of distances, we want to minimize their product. We cannot multiply variables with a linear solver, but we can slightly transform the problem by taking the logarithms of the distances and minimizing the sum of the logs.

- Alternatively, we might be interested in the longest path between start and end. In theory, this is unlikely to be solved by linear programming in all cases[7] but the pathological cases that hinder the theory are few and may not apply to the problem at hand. Another way to view this is that there is a large class of networks where it is possible to find longest paths.

---

[7]Note that if we can solve the longest path, we can solve the Hamiltonian path. Also, recall that linear programs can be solved in polynomial time. Ergo, if we can solve the longest path problem via LP, we prove P = NP. Then, we collect one million dollars from the Clay Mathematics Institute.

The simplest transformation would be to change the minimization to a maximization. We can obtain a maximization by negating the distance matrix. But this will allow repeated nodes; worse, it may lead to an unbounded model (an infinite loop). The problem is that the "flow" could go around a cycle an infinite number of times. A partial solution is to add constraints to ensure that no more than one unit of flow goes into any node. This is a redundant constraint in the case of a minimization, but not in the case of a maximization. In this manner we get rid of the infinite loop and the repeated nodes. There still remains a problem: subtours. I will explain and handle this problem in Section 5.4 of Chapter 5.

In the case of a cycle-free directed graph, then longest paths are easy to find. You saw it before a situation where these paths are of interest: in Section 2.3.1 of Chapter 2 I discussed *critical paths* of project management. These are the sequence of tasks which, if delayed, will delay the whole project. Note that these paths are rarely unique, so that looking for *a* (or worse, *the*) longest path is misguided (for a simple example, see Tasks 1 and 2 Figure fig:process-example). Let's create a small function that will start from the optimal solution of our project management model and use our shortest paths model to extract critical paths. See Listing 4-5 for details.

Listing 4-5. Critical Tasks Extractor

```
1    def critical_tasks(D,t):
2      s = set([t[i]+D[i][1] \
3            for i in range(len(t))]+[t[i] for i in
             range(len(t))])
4      n,ix,start,end,times = len(s),0,min(s),max(s),{}
5      for e in s:
6        times[e]=ix
7        ix += 1
8      M = [[0 for _ in range(n)] for _ in range(n)]
9      for i in range(len(t)):
```

```
10        M[times[t[i]]][times[t[i]+D[i][1]]] = -D[i][1]
11     rc,v,Path,Cost,Cumul = solve_model(M,times[start],
       times[end])
12     T = [i for i in range(len(t)) \
13         for time in Path if times[t[i]+D[i][1]] == time]
14     return rc, T
```

The first few lines create a set with all the tasks' starting and ending times; they will become our network nodes once we rename them 0, … , $n - 1$. At line 9 we create our distance matrix by the negative of the duration of each task. There is an entry per task, from its starting to ending time.

We then call our shortest path model, which in this case will find a longest path from earliest time to project completion time. Finally, we extract all the tasks that end on one of the nodes of the longest path. All these tasks are critical since they will stretch the already longest path if delayed.

Running this code on the example in Table 2-8 produces Table 4-10.

Table 4-10. *Critical Tasks of Project Management Example*

[0 1 2 6 7 9]

- We might be interested in the shortest paths tree from a start node to every other node in the network. In this case, we could run our shortest path model $n - 1$ times, but it is simple and interesting to create a separate model, especially since we can return the solution in a more compact form than $n - 1$ lists of paths.

  The idea of Listing 4-6 is to set the starting node with a supply of $n - 1$ (at line 8) and every other node with a demand of one (at line 12). The decision variables at line 5 each have an empty range if there is no corresponding arc or else a range of [0, n]. This is in contrast to the previous shortest path code where the range was up to one. We need this relaxed range because the flow will not be a unit flow until the very last arc on a given path; that is, the arc incident to a leaf.

We return a list of arcs in the tree, along with their distances as in Table 4-11. This is best displayed graphically as in Figure 4-7.

Listing 4-6. Shortest Paths Tree Model

```
1   def solve_tree_model(D,Start=None):
2     s,n = newSolver('Shortestupathsutreeuproblem'),len(D)
3     Start = 0 if Start is None else Start
4     G = [[s.NumVar(0,0 if D[i][j] is None else min(n,D[i]
      [j]),")\
5     for j in range(n)] for i in range(n)]
6     for i in range(n):
7       if i == Start:
8         s.Add(n-1 == sum(G[Start][j] for j in range(n)))
9         s.Add(0 == sum(G[j][Start] for j in range(n)))
10      else:
11        s.Add(sum(G[j][i] for j in range(n)) - \
12            sum(G[i][j] for j in range(n))==1)
13    s.Minimize(s.Sum(G[i][j]*(0 if D[i][j] is None else D[i]
      [j]) \
14                    for i in range(n) for j in range(n)))
15    rc  = s.Solve()
16    Tree = [[i,j, D[i][j]] for i in range(n) for j in range(n) \
16          if SolVal(G[i][j])>0]
17    return rc,ObjVal(s),Tree
```

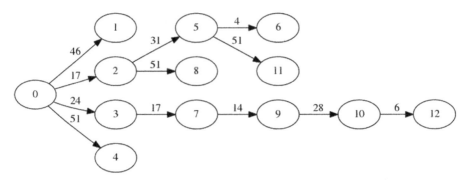

Figure 4-7. *Shortest tree solution*

Table 4-11. *Optimal Solution to the Shortest Paths Tree Problem*

From	To	Distance
0	1	46
0	2	17
0	3	24
0	4	51
2	5	31
2	8	51
3	7	17
5	6	4
5	11	51
7	9	14
9	10	28
10	12	6

- We might also be interested in the shortest paths between each pair of nodes. Again, if the reader is well-versed in combinatorial algorithms, he might be aware of the Floyd-Warshall algorithm, but for the same reasons we create a shortest path model, we might create an all-pairs shortest paths model or use our current model repeatedly to find all pairs. You will see later a better approach but since a few lines of code will suffice, let's write an all-pairs function (Listing 4-7).

To avoid running n_2 instances we use the *Principle of Optimality*, which states that, if $P = (v_{i+1}, v_{i+2}, v_{i+3}, \dots, v_{i+k})$ is a shortest path, then so is every subpath of P. This is used in the loop starting at line 11 to extract all intermediate paths from a given shortest path.

Listing 4-7. All-Pairs Shortest Path Function Using Our Shortest Path Model

```
1    def solve_all_pairs(D):
2      n = len(D)
3      Costs =[[None if i != j else 0 for i in range(n)]\
4          for j in range(n)]
5      Paths =[[None for i in range(n)] for j in range(n)]
6      for start in range(n):
7        for end in range(n):
8          if start != end and Costs[start][end] is None:
9            rc, Value, Path, Cost, Cumul = solve_model(D,start,end)
10           if rc==0:
11             for k in range(len(Path)-1):
12               for l in range(k+1,len(Path)):
13                 if Costs[Path[k]][Path[l]] is None:
14                   Costs[Path[k]][Path[l]] = Cumul[l]-Cumul[k]
15                   Paths[Path[k]][Path[l]] = Path[k:l+1]
16     return Paths, Costs
```

Running Listing 4-7 on our example produces the matrix of distances at Table 4-12. The reader should notice that this matrix extends the initial distance of matrix of Table 4-7.

Table 4-12. *Optimal Solution to All-Pairs Shortest Paths Problem*

	P0	P1	P2	P3	P4	P5	P6	P7	P8	P9	P10	P11	P12
P0		46	17	24	51	48	52	41	68	55	83	99	89
P1	46		63	70	31	33	37	54	49	68	90	81	96
P2	79	38		68	34	31	35	58	51	66	88	82	94
P3	24	70	41		33	37	41	17	49	31	59	81	65
P4	51	85	57	41		4	8	31	18	39	60	50	66
P5	48	81	53	37	4		4	27	22	35	57	51	63
P6	52	86	58	33	1	5		32	19	40	59	51	65
P7	75	54	26	64	31	27	31		49	14	42	66	48
P8	68	89	51	49	18	20	17	43		55	57	32	63
P9	83	68	40	72	39	35	39	14	57		28	80	34
P10	111	145	116	101	60	64	68	91	65	99		58	6
P11	100	121	83	81	50	52	49	75	32	61	58		56
P12	127	148	110	108	77	79	76	102	59	114	114	56	

CHAPTER 5

Classic Discrete Models

The problems in this chapter are classical examples of *integer programs* (IP). A better name would be *discrete linear programs* because we described the past ones as continuous linear programs and the antonym of continuous is discrete. Alas, the tradition is firmly entrenched so we will refer to them as IPs. They are characterized by algebraically linear constraints and linear objectives with the additional requirement that variables must take on only integral values.

All are very simple to state, if not always simple to model or solve. They are included here to highlight two elements:

- First, many real-life problems have embedded in them one or more of these simple, *pure* problems. It is therefore profitable for a modeler to recognize these kernels and model them with ease.

- Second, many of the problems, to be efficiently modeled, require some trickery. Knowing these tricks, and recognizing different situations where they can be applied, is the hallmark of a good modeler.

What makes an integral model is the requirement that some or all of the variables be integral. Keep in mind that, contrary to the pseudo-integral models of the previous chapter, the structure of the

© Serge Kruk 2018
S. Kruk, *Practical Python AI Projects*, https://doi.org/10.1007/978-1-4842-3423-5_5

problem does not guarantee this integrality and the modeler must choose a solver capable of handling integrality constraints.

There are a few reasons to require integral variables. The first and most obvious case is that we are counting objects, not measuring amounts (people, cars, or planets as opposed to water, carbon dioxide, or percentages). The second case occurs when the decision variables represent answers to yes/no questions (Should we build this plant? Should we get married?) or, more generally, Boolean conditions (with states of either True or False, satisfying the principle of an excluded middle). The third case more technical applies to auxiliary variables when they are used as "indicator variables." This is when they indicate the presence or absence of a certain state (y is 1 if and only if the continuous variable x is non-zero). Of course, the boundary between these use cases is blurry: a true decision variable could be an indicator variable and an auxiliary variable could be counting people. The three cases are nevertheless good to keep in mind while modeling.

Problems of this chapter have multiple interesting variations. I cannot possibly hope to cover them all, but the reader, after reading some of the variations, is encouraged to imagine others. No matter how creative one is in varying some of the requirements, most of these problems have been studied so extensively that few variations remain untouched and most have found some use.[1]

5.1 Minimum Set Cover

The first problem in this chapter is one of the most studied and best understood of the integer programs. There are a number of applications, but let's consider the following one: General Engine Corp. is considering

---

[1]This might be a case, in the words of Wigner, of the unreasonable effectiveness of mathematics in the natural sciences; or more prosaically, because we mostly solve problems that we know how to solve.

suppliers for its new line of electric cars. Every supplier can produce some parts of the cars, and there is overlap in the parts coverage between suppliers. For instance, Dolphin Inc. can supply wheel bearings, electrical cables, and low-power light emitting diodes while Schukert GA can supply electrical cables, batteries, and battery casings. There are hundreds of suppliers and thousands of parts.

For General Engine, minimizing the number of suppliers offers contractual savings. So the goal is to find the smallest number of suppliers that together will provide all the required parts. The name *set cover* is explained by the goal: covering all the elements of the set, here the parts needed to build the electric cars. The small example used to illustrate the model is given in Table 5-1.

Table 5-1. *Example of Set Cover*

Supplier	Part Numbers	Supplier	Part Numbers
S0	{ 3; 4; 5; 8; 24 }	S1	{ 11; 15; 21; 23 }
S2	{ 9; 15; 24 }	S3	{ 9; 13 }
S4	{ 5; 11; 12; 14; 16; 20 }	S5	{ 8; 11; 12; 15; 21 }
S6	{ 1; 4; 18; 20 }	S7	{ 0; 3; 6; 11; 13; 15; 21; 23 }
S8	{ 14; 16; 18; 19; 23 }	S9	{ 2; 7; 16; 22 }
S10	{ 10; 14; 21 }	S11	{ 6; 19 }
S12	{ 4; 10; 24 }	S13	{ 3; 4; 7; 9; 17 }
S14	{ 1; 3; 5; 6; 15; 18; 19; 20; 23 }		

5.1.1 Constructing a Model

This model will be described in stages.

5.1.1.1 Decision Variables

What we need to decide in this problem is which suppliers will get contracts. This is a yes/no decision. We need, for each supplier, a variable that will take on one of two values. The classical approach in integer programming is to use an integer variable with a range of $[0, 1]$. Being an integer, it therefore only has two possible values, zero and one, and is known as a *binary* or *indicator* variable.[2]

There are other possible approaches: one using Boolean variables taking on values `True` and `False`, but this is really the same approach, renamed; and one using a dynamic array variable that will include only the chosen suppliers. This later approach may seem natural at first glance, but is not easily implemented using an integer solver. It is better suited to a constraint solver, which I will not cover here.

Let's assume a set S of suppliers and declare our first integer variables as

$$S_i \in \{0,1\} \ \forall i \in S$$

The interpretation is that if, for example, s_3, s_5, and s_7 are one while all the others are zero, then General Engine awards a contract to suppliers 3, 5, and 7 only. It may occur to the reader that in cases where the number of suppliers is much larger than the final set of chosen suppliers we are wasting resources. We will try to mitigate this waste but in a certain sense, it is unavoidable in integer programs.

---

[2]There are cases where a binary choice -1, 1 would make the model simpler. Alas, no popular integer solver offers that option.

5.1.1.2 Objective

The standard objective is to minimize the number of suppliers. Since we have a zero-one variable per supplier, we need to minimize the sum of all these, so

$$\min \sum_{i \in S} S_i$$

It is possible to encounter a cost per supplier. So that, instead of simply minimizing the number of suppliers, we want to minimize the total cost. Assuming that we have cost C_i for supplier i, we modify the objective function to read

$$\min \sum_{i \in S} C_i S_i$$

This cost array could be a function of the parts supplied (more parts, larger cost) and the bargaining strength of the supplier.

5.1.1.3 Constraints

From a high-level perspective, there is only one constraint: General Engine must have access to all the required parts. Of course, there may be more than one supplier for some parts, but what we must not have is a part with no supplier (a car without a steering wheel might not sell so well).

How can we insure that we have all of the parts? Consider a given part, say part 23. Which suppliers provide it? There may be four, say 1, 7, 8, and 14.

This means that we must choose one of these suppliers to get part 23 or, algebraically, that the sum $s_1 + s_7 + s_8 + s_{14}$ must be at least one. (Not equal to one because there are, in general, no solutions without some redundancy.)

This leads us to a constraint per part j in the set P of all parts. We will assume that we have sets P_i of parts supplied by supplier i, just as in Table 5-1, to get

$$\sum_{i:j\in P_i} s_i \geq 1 \; j \in P \tag{5.1}$$

The notation $\{i : j \in P_i\}$ is meant to indicate that we choose index i only if index j is in the set P_i. We will see how easily this is accomplished in the executable model.

5.1.1.4 Executable Model

In Listing 5-1, we see the whole model. Let's look at it carefully, highlighting the two major differences with all the previous models:

- The solver instantiation

- The decision variable declaration

Listing 5-1. Set Cover Model (set cover.py)

```
1   def solve_model(D,C=None):
2       t = 'SetuCover'
3       s = pywraplp.Solver.CBC_MIXED_INTEGER_PROGRAMMING
4       s = pywraplp.Solver(t,s)
5       nbSup = len(D)
6       nbParts = max([e for d in D for e in d])+1
7       S = [s.IntVar(0,1,") for i in range(nbSup)]
8       for j in range(nbParts):
9         s.Add(1 <= sum(S[i] for i in range(nbSup) if j in D[i]))
10        s.Minimize(s.Sum(S[i]*(1 if C is None else C[i]) \
11          for i in range(nbSup)))
12        rc = s.Solve()
13        Suppliers = [i for i in range(nbSup) if SolVal(S[i])>0]
```

```
14    Parts = [[i for i in range(nbSup) \
15       if j in D[i] and SolVal(S[i])>0] for j in
         range(nbParts)]
16    return rc,ObjVal(s),Suppliers,Parts
```

The function receives two-dimensional array D containing the part numbers supplied by each supplier, exactly as in Table 5-1. The code also accepts cost array C in case there is a varying cost per supplier. It is optional and its absence indicates a pure set cover problem, one where we are concerned with minimizing the number of subsets chosen.

Line 4 is different from all our previous models. It chooses a solver, in this case, CBC from the COIN-OR project,[3] which can handle discrete as well as continuous variables. This very small change on our part represents an order of magnitude change on the part of the solver. In fact, to solve an integer model, most solvers will internally solve a multitude of continuous models derived from ours. The algorithms are fascinating but beyond the scope of this book.[4]

For this first discrete model, we create the solver instance using the low-level OR-Tools routine `Solver`. From here onward, we will use our own `newSolver` in this manner:

```
s = newSolver('Name of problem', True)
```

The second parameter, which defaults to `False`, instantiates an integer solver if `True`. Internally we usually use CBC, but there are a number of possible integer solvers. (See Listing 7-31 in Chapter 7 for details.)

Line 7 defines our binary variable (with the understanding that zero will mean "ignore supplier" and one will mean "pick supplier"). Up until now, all variables were defined with `NumVar`, which implies a floating point

---

[3]`www.coin-or.org`

[4]The interested reader should search "branch and bound" to start reading about the solution techniques.

variable approximating a real number. With `IntVar` we are instructing the solver that a variable can only take on integral values. Since we give it a range of zero to one, it forces the variable to have only one of two values. Any range is possible with all solvers.

The reader should experiment with this model by changing the `IntVar` to a `NumVar` and note that the variables will now take on values of zero, one-half, and one.[5] What does it mean to have one-half of a supplier? Nothing, hence the integrality requirement.

The loop at line 9 implements the cover constraints. It mimics to the letter the constraint (5.1), forcing the sum of suppliers of each part to be above one. Notice how easily we can extract subsets based on conditionals in Python.

The cost function at Line 11 is either the number of supplier, as in a traditional set cover, or the total cost of choosing these suppliers if each one incurs a different cost. This is the purpose of the optional cost array C, indexed by supplier.

Finally, after we solve it, we construct meaningful return values. It would be painful to the caller to receive the raw S variables. Most of them might be zero. In a real problem with thousands, maybe tens of thousands of parts and suppliers, the zeros are not interesting. So we return an array containing only the suppliers who should be offered a contract, along with a cross-reference of parts to suppliers. This way the user knows where to go for each part.

For our example, the solution, absent a cost array, is displayed in Table 5-2. The first line lists all retained suppliers, the next indicates who can supply (among those retained) each part. Note that each part is covered.

---

[5]No other fractions are possible in a pure set cover problem for fascinating reasons the reader is encouraged to research (keyword search: half-integrality).

Table 5-2. *Optimal Solution to the Set Cover Problem*

Parts	Suppliers	Parts	Suppliers
All	{ 5; 7; 9; 10; 12; 13; 14 }	Part #0	{ 7 }
Part #1	{ 14 }	Part #2	{ 9 }
Part #3	{ 7; 13; 14 }	Part #4	{ 12; 13 }
Part #5	{ 14 }	Part #6	{ 7; 14 }
Part #7	{ 9; 13 }	Part #8	{ 5 }
Part #9	{ 13 }	Part #10	{ 10; 12 }
Part #11	{ 5; 7 }	Part #12	{ 5 }
Part #13	{ 7 }	Part #14	{ 10 }
Part #15	{ 5; 7; 14 }	Part #16	{ 9 }
Part #17	{ 13 }	Part #18	{ 14 }
Part #19	{ 14 }	Part #20	{ 14 }
Part #21	{ 5; 7; 10 }	Part #22	{ 9 }
Part #23	{ 7; 14 }	Part #24	{ 12 }

5.1.2 Variations

Variations abound, from seemingly unrelated fields.

- A famous instance of the set cover problem is part of the infamous *crew scheduling* problem. Imagine that we are an airline and we want to make sure that all the so-called *legs* (pairs of cities) are covered during a certain time window. We have rosters of crew members travelling together from city A to city B, with stopovers in cities C, D, ... , E. Our task is to cover all legs using the minimum number of rosters.

133

- Another geeky example involves computer virus detection. Imagine that we have a database of thousands of computer viruses and are trying to build a detector. One way to do this is to try to identify short strings of bytes that are present in these viruses but not in non-virus code. What we want is to minimize the numbers of strings and yet identify all viruses. Then our detector will look for this small set of strings in the data (all programs on the hard drive).

- There are applications in telecommunications. Imagine that we can build cell towers in a city in a number of locations. Considering the cost of each, we want to minimize expenditures and yet cover all the buildings and houses in the city.

- Where should we locate fire stations in a city so that, considering the average response time, we minimize the number of stations and yet cover the whole city?

5.2 Set Packing

The mirror problem to the set cover is known as the set packing. In either case, we are given a universal set and a set of subsets, and we need to choose some of them. In the former case, we aim at covering the universal set with a minimal set of subsets, possibly covering some elements more than once. In the latter case, the objective is to choose as many of the subsets as possible but without ever choosing an element more than once. Therefore, some elements may not be covered.

To justify this problem, let's consider an application for airline crew scheduling. To simplify, say that each plane must have a pilot, a co-pilot, a navigator, and a purser. Each of these sets is called a roster.

Some pilots may fly some types of planes but not others. Pilots also may have preferences for their co-pilots (and vice versa). Conceptually we can think of a specific combination of plane/pilot/co-pilot/navigator/purser as a subset of our universal set of planes and crews members. What we want is to maximize the number of subsets we choose, but we must not pick two subsets that share elements since a pilot cannot be at two places at once. Table 5-3 illustrates a small instance of this problem.

Table 5-3. *Example of Set Packing from Crew Scheduling*

Roster #	Crew IDs	Roster #	Crew IDs
0	{ 3; 18; 30 }	1	{ 4; 4; 36 }
2	{ 1; 5; 9 }	3	{ 7; 17; 30 }
4	{ 10; 23; 23 }	5	{ 8; 10; 25 }
6	{ 19; 29; 36 }	7	{ 3; 4; 17 }
8	{ 19; 28; 40 }	9	{ 11; 24; 31 }
10	{ 18; 30; 33 }	11	{ 22; 25; 26 }
12	{ 13; 15; 26 }	13	{ 21; 27; 28 }
14	{ 7; 12; 33 }		

5.2.1 Constructing a Model

The model will be constructed in stages.

5.2.1.1 Decision Variables

What we need to decide in this problem is very similar to the decision for the set cover: which rosters to pick. Again, it's a yes/no decision, which suggests an indicator variable. Let's assume a set S of crew rosters to declare our indicator variables as

$$s_i \in \{0,1\} \ \forall i \in S$$

5.2.1.2 Objective

The simplest objective is to maximize the number of rosters chosen, therefore

$$\max \sum_{i \in S} S_i$$

Of course, we could also have a variation with a value per roster and maximize the total value.

5.2.1.3 Constraints

The constraint, and there is only one, is never to pick two rosters including the same crew member. Since our decision variables are zero-one, we can simply force, for each crew, that the sum of roster variables including the crew under consideration is at most one.

If all crew of roster i are held in Si and the universal set of crew is U, we obtain

$$\sum_{i:j \in S_i} s_i \leq 1 \ \forall j \in u$$

5.2.1.4 Executable Model

The executable model is seen in Listing 5-2. Very similar to Listing 5-1, it receives a two-dimensional array D with a list of crew rosters, exactly as Table 5-3. The function will also accept an optional cost array C to attach to each roster.

Listing 5-2. Set Packing Model (set packing.py)

```
1  def solve_model(D,C=None):
2      s = newSolver('SetuPacking', True)
3      nbRosters,nbCrew = len(D),max([e for d in D for e in d])+1
4      S = [s.IntVar(0,1,") for i in range(nbRosters)]
```

```
 5    for j in range(nbCrew):
 6      s.Add(1 >= sum(S[i] for i in range(nbRosters) if j in D[i]))
 7    s.Maximize(s.Sum(S[i]*(1 if C==None else C[i]) \
 8      for i in range(nbRosters)))
 9    rc = s.Solve()
10    Rosters=[i for i in range(nbRosters)if S[i].
      SolutionValue()>0]
11    return rc,s.Objective().Value(),Rosters
```

A solution for our instance, with no cost array, appears in Table 5-4.

Table 5-4. *Optimal Solution to the Set Packing*

Rosters chosen 8	{ 2; 4; 6; 7; 9; 12; 13; 14 }

5.2.2 Variations

A few minor variations.

- The main variation is to have a cost on the rosters selected. We then minimize the total cost. The code given already implements this possibility.

- Another possibility is that we have a combination of set cover and set packing: we want to cover entirely the universal set and use each element exactly once. In this case, we speak of set partitioning.

5.3 Bin Packing

Notwithstanding the familial appearance of set packing and the current problem, *bin packing*, the two problems are antipodal in difficulty. Set cover is the easy cousin while set packing is the battle-hardened old aunt. To solve it, we will need to go past the obvious natural formulation and dig into a new bag of tools.

Abstractly, bin packing is the problem of partitioning a set, where each element has a weight so that we minimize the number of groups and yet maintain each group under a prescribed weight limit.

To illustrate, shipper VQT Inc. has a number of trucks, each with a maximum weight capacity. On a particular morning, they have packages of various weights to transport. A simple instance is described in Table 5-5. The goal is to minimize the number of trucks used to deliver all packages.

***Table* 5-5.** *Example of Bin Packing*

	Truck Weight Limit Number of Packages	1264 Unit weight
0	8	258
1	10	478
2	8	399
Total	26	10036

Note that having only a weight limit is not entirely realistic. Packages also have a volume and it is likely that we need to be able to pack according to volume. But that problem is considerably more difficult and we will leave it aside. There should also be consideration of distances, which we will tackle in a later section (see 5.4). To repeat a point worth repeating: few, if any, real-life optimization problems are pure and simple textbook problems. They are always a combination of multiple problems. A good modeler recognizes this and has the toolset to model all.

5.3.1 Constructing a Model

The model will be described in stages.

5.3.1.1 Decision Variables

What we need to decide is, "Which package goes into which truck?" We know all the packages. Although there are many instances of identical packages (identical for our purposes, that is, with the same weight), we can give them ordinal numbers. But we do not really know the number of trucks.

That is one of the questions we are trying to answer. Nevertheless, we can certainly give an upper bound on the number of trucks by some heuristic. At worst, we can say with certainty that we will need at most one truck per package.

So let's assume P packages and at most T trucks. Our decision variable is then

$$x_{i,j} \in \{0,1\} \ \forall i \in P, j \in T$$

where $x_{i,j} = 1$ will mean that package i goes into truck j.

This is a good start but we also need to know which of the trucks we will need. So another decision variable seems indicated, namely

$$y_j \in \{0,1\} \ \forall j \in T$$

where $y_j = 1$ will indicate that truck j is to be used. This seems to answer all of the questions we need answered.

5.3.1.2 Constraints

First, we need to establish a relationship between our $x_{i,j}$ and y_j variables since we must have a given y_j equal to one (truck j used) if $x_{i,j}$ is one for any i. Another way to view this is that we cannot put packages in a truck we do not use. You have seen this type of constraint before. Recall the diet problem of Section 2.1.2 in Chapter 2, where one of the constraints was "If food 2 is used, then we must have at least as much food 3 in the diet."

The general idea is to ensure that one variable is bounded by another or by a multiple of another. In this case, that trick suggests using

$$x_{i,j} \leq y_i \ \forall i \in P, \forall j \in T \tag{5.2}$$

This does satisfy our relationship, although it may look rather wasteful to the reader. Indeed, we will prune this attempt shortly. Note coincidentally that we also need to bind the sum of the package weights in every certain truck. Assuming that package i has weight w_i and truck j has capacity W_j, we need

$$\sum_{i \in P} w_i x_{i,j} \leq W_j \ \forall j \in T \tag{5.3}$$

Here is the obvious question now, once we notice the explosion of similar constraints: "Is there a way to combine equations (5.2) and (5.3)?" Indeed there is:

$$\sum_{i \in P} w_i x_{i,j} \leq W_j y_j \ \forall j \in T \tag{5.4}$$

We can see that equation (5.4) subsumes both of (5.2) and (5.3). We have reduced the number of constraints from $|P||T| + |T|$ to $|T|$, a non-trivial improvement.

At this point, our model guarantees that

- A truck is used if any package is loaded in it.

- The sum of the package weights in a truck respects its capacity.

Last, we ensure that each package finds its way to some truck:

$$\sum_{i \in T} x_{i,j} = 1 \ \forall i \in P \tag{5.5}$$

5.3.1.3 Objective

The simplest objective is to minimize the number of trucks used, therefore

$$\min \sum_{j \in T} y_j$$

5.3.1.4 Executable Model

We will assume that the function receives array D containing a list of packages with their weights and the count of packages of each weight (we will call these weight classes), exactly as in Table 5-5. It also receives a weight capacity for each truck in W. The third parameter, optional, will be explained after we solve our small example. Just note that its default value is False and in that case, a large set of, as yet unexplained, constraints are skipped (lines 17 to 27). See Listing 5-3.

Listing 5-3. Bin Packing Model (bin packing.py)

```
1   def solve_model(D,W,symmetry_break=False,knapsack=True):
2     s = newSolver('BinuPacking',True)
3     nbC,nbP = len(D),sum([P[0] for P in D])
4     w = [e for sub in [[d[1]]*d[0] for d in D] for e in sub]
5     nbT,nbTmin = bound_trucks(w,W)
6     x = [[[s.IntVar(0,1,") for _ in range(nbT)] \
7        for _ in range(d[0])] for d in D]
8     y = [s.IntVar(0,1,") for _ in  range(nbT)]
9     for k in range(nbT):
10      sxk = sum(D[i][1]*x[i][j][k] \
11              for i in range(nbC) for j in range(D[i][0]))
12      s.Add(sxk <= W*y[k])
13     for i in range(nbC):
14      for j in range(D[i][0]):
15        s.Add(sum([x[i][j][k] for k in range(nbT)]) == 1)
```

141

```
16    if symmetry_break:
17      for k in range(nbT-1):
18        s.Add(y[k] >= y[k+1])
19      for i in range(nbC):
20        for j in range(D[i][0]):
21          for k in range(nbT):
22            for jj in range(max(0,j-1),j):
23              s.Add(sum(x[i][jj][kk] \
24                for kk in range(k+1)) >= x[i][j][k])
25            for jj in range(j+1,min(j+2,D[i][0])):
26              s.Add(sum(x[i][jj][kk] \
27                  for kk in range(k,nbT))>=x[i][j][k])
28    if knapsack:
29      s.Add(sum(W*y[i] for i in range(nbT)) >= sum(w))
30    s.Add(sum(y[k] for k in range(nbT)) >= nbTmin)
31    s.Minimize(sum(y[k] for k in range(nbT)))
32    rc = s.Solve()
33    P2T=[[D[i][1], [k for j in range(D[i][0]) for k in
      range(nbT)
34                  if SolVal(x[i][j][k])>0]] for i in range(nbC) ]
35    T2P=[[k, [(i,j,D[i][1]) \
36      for i in range(nbC) for j in range(D[i][0])\
37            if SolVal(x[i][j][k])>0]] for k in range(nbT)]
38    return rc,ObjVal(s),P2T,T2P
```

At line 4 we construct an array of weights, one per package. This implicitly also assigns an ordinal to each package. The function bound-trucks, described in Listing 5-4, uses the weights of the packages and the capacity of each truck to quickly estimate an upper bound on the number of trucks. This function does not need to be brilliant, but a better bound tends to accelerate the solver.

The two lines starting at 7 define our decision variables: one to assign packages to trucks, and one to select trucks. The package variable is a three-dimensional array. The first dimension indicates the weight class, the second is the ordinal within the class, and the third is the truck. So that if, for example, *x[2][3][5]* has value of 1, it will mean that package three of the weight class two is loaded onto truck five.

The constraint at the loop on line 9 is a transcription of equation (5.4), our *merged* constraint to both force the truck selection variable and to limit the total package weight carried by a truck, modified to use the three-dimensional decision variables.

The final constraint at line 15 is a transcription of equation (5.5) to ensure all packages find a truck.

Ignore the lines starting at 16 and guarded by the `symmetry_break` parameter, for now.

After solving, we produce two arrays, each providing a distinct view of the solution. The first indicates, for each package, the loading truck. The second indicates, for each truck, the list of packages. The solution for our instance appears in Table 5-6, `tab:bin_packing_results_bad`. The first table lists the trucks and their content indicated by a triple (weight class, package ordinal, weight). The second table lists each weight class, in the same order as Table 5-5, with the truck in which each package of the class is loaded.

Table 5-6. *Optimal Package Assignments (Naive Approach)*

Weight	Truck Id
258	[0, 6, 2, 5, 3, 8, 7, 4]
478	[3, 5, 6, 8, 4, 5, 4, 6, 7]
399	[2, 0, 7, 8, 0, 3, 2]

Even for small instances, this code can take hours to produce a solution. Bin packing is not an easy problem and here is part of the reason why: notice that some truck numbers are omitted. The solver seems to choose trucks at random among those we allow. Also, the packages of a given weight class are randomly distributed among the trucks. Indeed, executing the same model on the same instance on a different computer or a different solver might very well produce a different answer (of course, with the same total number of trucks). The problem is that there are many solutions with exactly the same value. Imagine, for instance, swapping the entire contents of two trucks with the same capacity or swapping two packages of the same weight class within a truck, or between two trucks. These swaps will clearly have no effect on the value of the solution.

In classical optimization terms, this situation is a form[6] of *degeneracy*. Researchers in constraint programming talk[7] of *symmetry*. It almost always affects the solver negatively. The runtime is difficult to predict because it is solver-dependent but it is rarely good. There is another reason to want to modify the model to avoid these identical solutions: we could produce nicer solutions for the user.

Adding constraints favoring one optimal solution over another identical one (identical from our point of view) is known as *symmetry breaking*, which begins to explain the parameter `symmetry_break` in the code guarding the additional constraints. Let's tackle these constraints, starting with the easiest first.

How can we ensure that the trucks are chosen in order and none are skipped, assuming that they all have the same capacity? One way is to bind the truck selection variables pairwise:

$$y_{j-1} \le y_j \ \ \forall j \in T \setminus \{0\}$$

To see how this works, consider what happens to the y vector for, say, y_5 to be one. It must be that y_4 is one and, transitively, so must be $y_3, y_2, y_1,$ and y_0. On the other hand, it has no effect on y_6 or higher. In terms of code, this is done in the loop at line 17.

---

[6]For the theoretically minded, this is a case of dual-degeneracy.

[7]The symmetry stems from visualizing the search tree and noticing that there are multiple branches with exactly the same structure and value.

The second form of symmetry alluded to is interchangeable packages. The way we stated the problem, there is no difference between two packages in the same weight class. Yet, for the solver, swapping two packages within a truck or between two trucks is another potential solution, and any time spent looking in that direction is time wasted.

Let's consider how to break this symmetry by looking at a small example, say three packages and three trucks. The idea is that since the packages of a weight class are naturally ordered, we can force that they be loaded into trucks in their order. For instance, if the second package is loaded in truck one, then the third can only be loaded in truck one or higher. In terms of the decision variables, we want the following implications:

$$x_{0,2} = 1 \Rightarrow x_{1,2} = 1 \wedge x_{2,2} = 1$$
$$x_{0,1} = 1 \Rightarrow x_{1,1} + x_{1,2} = 1 \wedge x_{2,1} + x_{2,2} = 1$$
$$x_{0,0} = 1 \Rightarrow x_{1,0} + x_{1,1} + x_{1,2} = 1 \wedge x_{2,0} + x_{2,1} + x_{2,2} = 1$$
$$x_{1,2} = 1 \Rightarrow x_{2,2} = 1 \qquad\qquad \wedge\, x_{0,0} + x_{0,1} + x_{0,2} = 1$$
$$x_{1,1} = 1 \Rightarrow x_{2,1} + x_{2,2} = 1 \qquad\qquad \wedge\, x_{0,0} + x_{0,1} = 1$$
$$x_{1,0} = 1 \Rightarrow x_{2,0} + x_{2,1} + x_{2,2} = 1 \qquad\qquad \wedge\, x_{0,0} = 1$$
$$x_{2,2} = 1 \Rightarrow \qquad x_{0,0} + x_{0,1} + x_{0,2} = 1 \wedge x_{1,0} + x_{1,1} + x_{1,2} = 1$$
$$x_{2,1} = 1 \Rightarrow \qquad\qquad x_{0,0} + x_{0,1} = 1 \wedge x_{1,0} + x_{1,1} = 1$$
$$x_{2,0} = 1 \Rightarrow \qquad\qquad\qquad x_{0,0} = 1 \wedge x_{1,0} = 1$$

You will see later (Section 7.2.3 in Chapter 7 on reification) a general way to implement these implications. For now, let's try to implement them as simply as possible.

The first implication says "If package 0 is loaded onto truck 2, then both packages 1 and 2 must also be loaded onto truck 2." But this is a boundary case because truck 2 is our last truck. The next implication is more interesting: "If package 0 is loaded onto truck 1, then packages 1 and 2 must be loaded onto truck 1 or 2."

In all generality, "If a package is loaded onto a given truck, then all packages with greater ordinal numbers must be loaded onto trucks of greater or equal ordinal numbers." Notice that the constraint structure is that if some variable takes on value one, we must have an equation *with right hand side one* holding. The unit *right hand side* is our ticket, as we can use the conditioning variable as the right-hand side.

Beware of over-constraining the model. Consider part of the second implication $x_{0,1} = 1 \Rightarrow x_{1,1} + x_{1,2} = 1$ and the following naive approach. If we use an equality, as in

$$x_{1,1} + x_{1,2} = x_{0,1}$$

the model will likely fail because if $x0,1$ is zero (for example if package 0 is loaded onto truck 0 instead of 1), then we are preventing package 1 to be loaded onto trucks 1 or 2, which would be acceptable solutions. So the right constraint is

$$x_{1,1} + x_{1,2} \geq x_{0,1}$$

The right-hand side at zero trivializes the constraints as all variables on the left-hand side are non-negative; at one, it forces a correct assignment to a higher numbered truck. In more abstract terms, we have an inequality because we are implementing a logical implication, not a logical equivalence (a $\Rightarrow$, not a $\Leftrightarrow$). These constraints are implemented at line 27 of model 5.3.

The right column of the implications can be read as "If package i is loaded onto truck k, then all packages with lower ordinal numbers must be loaded onto trucks with lower or equal ordinal numbers." These constraints are mostly redundant, yet they can help some solvers under some conditions. The reader is encouraged to enable or disable some symmetry-breaking constraints and experiment.

All these additional constraints reduce the search space. With some solvers, on some problems the approach will drastically reduce the execution time. With these symmetry-breaking constraints enabled by calling solve_model with the last parameter as True, the output on the same instance is shown in Tables 5-7 and 5-8. You see that all trucks used are consecutive from zero and that the packages are loaded in order, a much nicer solution. What you do not see by reading the table is that the runtime is a very small fraction of the runtime necessary for the same instance when the symmetry-breaking constraints are ignored.

Table 5-7. *Optimal Truck Loads with Symmetry-Breaking Constraints*

Trucks 8.0 (Id Weight)	Packages 24 (9159) (Id Weight)*
0 (1252)	[(0, 0, 258), (0, 1, 258), (0, 2, 258), (1, 0, 478)]
1 (1214)	[(0, 3, 258), (1, 1, 478), (1, 2, 478)]
2 (1214)	[(0, 4, 258), (1, 3, 478), (1, 4, 478)]
3 (1135)	[(0, 5, 258), (1, 5, 478), (2, 0, 399)]
4 (1056)	[(0, 6, 258), (2, 1, 399), (2, 2, 399)]
5 (956)	[(1, 6, 478), (1, 7, 478)]
6 (1197)	[(2, 3, 399), (2, 4, 399), (2, 5, 399)]
7 (1135)	[(0, 7, 258), (1, 8, 478), (2, 6, 399)]

Table 5-8. *Optimal Package Assignments with Symmetry-Breaking Constraints*

Weight	Truck Id
258	[0, 0, 0, 1, 2, 3, 4, 7]
478	[0, 1, 1, 2, 2, 3, 5, 5, 7]
399	[3, 4, 4, 6, 6, 6, 7]

We have left to consider a simple heuristic to bound the number of trucks required. We start by adding packages to the first truck until we reach capacity and then move on to the next. This greedy approach will never be optimal but is enough to get a reasonable upper bound on the required number of trucks. A simple lower bound is obtained by dividing the sum of the weights of all packages by the truck capacity. This is presented in Listing 5-4. Better heuristics abound and they might be necessary for large instances.

Listing 5-4. Simple-Minded Heuristic to Bound the Number of Trucks

```
1  def bound_trucks(w,W):
2    nb,tot = 1,0
3    for i in range(len(w)):
4      if tot+w[i] < W:
5        tot += w[i]
6      else:
7        tot = w[i]
8        nb = nb+1
9    return nb,ceil(sum(w)/W)
```

5.3.1.5 Variations

This problem often appears in combination with others. But here are a few of the simpler variations.

- It may be that each truck has a different weight capacity. The capacity constraint is simple to adapt, but care must be taken with the symmetry-breaking constraint. Skipping some trucks may be unavoidable. So the symmetry-breaking must be done only within subsets of trucks with the same capacity.

- Instead of loading a fixed number of packages in an undetermined number of trucks, we may have a fixed number of trucks and an undetermined number of packages to load. In that situation, packages, in addition to a weight, usually also have a value, and we must try to maximize the total value. This situation is usually much simpler to solve. Assuming that package i has value v_i, the objective function is

$$\max \sum_{i \in P} \sum_{j \in T} v_i x_{i,j}$$

subject to constraint (5.4).

- There is a simpler version of bin packing known as *knapsack* where packages have value and weight, but there is only one truck with a weight capacity. This problem is so simple that there exist very fast algorithms for it. But, a general-purpose integer solver will, of course, solve it without any difficulty. Although it is simple, it does have some value, not as a problem that occurs naturally, but as a subproblem of a more complex situation. You will see examples of this later.

- A closely related problem is that of *capital budgeting*. Consider a multi-period planning horizon T and a set of possible projects, P; each project j requiring an investment of a_{tj} in period t and representing a value c_j. Given a limited budget b_t in period t, which projects should be earmarked for investments? The model is a simplification of bin packing:

$$\max \sum_{j \in P} c_j y_j$$
$$\sum_{j \in P} a_{tj} y_i \le b_t \quad \forall t \in T$$
$$y_i \in \{0,1\}$$

where y_j represents the decision "Go ahead (or not) with project j."

5.4 TSP

We now tackle the venerable *travelling salesman problem* (hereafter TSP). This problem was never important for salesmen but it is very important in vehicle routing, electronic circuit design, and job sequencing, among other applications. Moreover, it will allow me to describe, with a minimum of spurious complexity, an effective and reusable modeling technique: adding constraints iteratively.

Here is an example situation: at HAL Inc., during the process of a new circuit design, power must be routed to each elementary component. These components are set in a two-dimensional lattice, potentially all pairwise connected. The best way to feed power to these components is to establish a path of minimal total length, conceptually starting at the power supply (V_{cc}), going around to each component, and then coming back to the power supply (Vee or ground).[8]

Therefore, the problem, viewed abstractly, can be stated as "Given a matrix of pairwise distances in a graph, find a tour of all vertices minimizing total distance." Table 5-9 is the example we will solve to illustrate. In addition to the distances, it includes the Cartesian coordinates of the points. We will not use these coordinates in the model but they are useful for visualizing the problem. The absence of a number in the table indicates that there is no direct path between the two nodes.

---

[8]Whether the trace comes back to the origin is irrelevant from a complexity standpoint. We can assume a distance zero between Vcc and Vee if need be.

Table 5-9. *Example of Distance Matrix for TSP*

P (x y)	P0	P1	P2	P3	P4	P5	P6	P7	P8	P9
P0 72 19		711	107	516	387	408	539	309	566	771
P1 10 37	539		769	881	380	546	655	443	295	1140
P2 77 31	122	752		281	441	264	318	448	588	730
P3 89 61	519	875	274		435	334	93	776	949	302
P4 51 61	484	561	338	419		118	268	607	495	431
P5 57 52	409	406	244	380	93		295	544	549	494
P6 82 69	479	735	334	101	345	247		679	809	238
P7 52 1	221	444	433	744	487	435	649		325	840
P8 21 14	510	303	599	984	531	553	847	350		1001
P9 88 96	663	989	664	335	588	434	297	1093	1012	

5.4.1 Constructing a Model

The model will be described in stages.

5.4.1.1 Decision Variables

What we need to decide in this problem is simply the path to take, which means the sequence of points to follow. This is identical to our decision in the shortest path problem; therefore, assuming that P is the set of points, we define

$$x_{i,j} \in \{0,1\} \ \forall i \in P, \forall j \in P,$$

where $x_{i,j}$ with a value of one will indicate that we need to connect points i and j. Beware: This problem has the same decision variable and underlying graph structure as the shortest path problem. However, it is not a flow problem; it is considerably more complex, as you will begin to appreciate shortly.

5.4.1.2 Objective

The objective function is identical to a shortest path model. Assuming that the distance matrix is D, we get

$$\min \sum_i \sum_j D_{i,j} x_{i,j}$$

5.4.1.3 Constraints

In the shortest path problem, when we entered an intermediate node, we had to ensure that we also exited it. Here we must ensure a tour, a single closed path covering every vertex exactly once. For each vertex, then, we must choose precisely one arc going in and one arc going out, as in

$$\sum_{j \in P \setminus \{i\}} x_{i,j} = 1 \ \forall i \in P \tag{5.6}$$

and

$$\sum_{j \in P \setminus \{i\}} x_{j,i} = 1 \ \forall i \in P \tag{5.7}$$

Now comes the real difficulty: distinguishing TSP from shortest path. Are the above two constraints enough? Perhaps surprisingly, no, they are not. Every vertex is on a tour, but there may be more than one such tour. Satisfying the above constraint, we could get a path 0, 1, 3, 4, 0, and another looping around the rest of the vertices. These problematic paths are known as subtours and must be eliminated. The key to the elimination is to realize that for any strict subset of the nodes, the number of chosen arcs must be less than the number of nodes. For instance, to eliminate the subtour 0, 1, 3, 4, 0, we could add the following constraint:

$$x_{0,1} + x_{1,0} + x_{1,3} + x_{3,1} + x_{3,4} + x_{4,3} + x_{4,0} + x_{0,4} \leq 3.$$

With the addition of this constraint, solvers will never include more than three arcs between the four problematic vertices, preventing a subtour among them. Another way to view this type of constraint: it forces a path entering a cluster of nodes to exit the cluster. The difficulty is that there are many possible subtours, in fact, an exponential number: every subset of vertices of size larger than one is a potential subtour.

Can we add all the possible subtours? Programmatically, this is not difficult, but the resulting model would be unwieldy and many solvers would slow down unacceptably. The trick is to improve the model iteratively, as we did when optimizing a non-linear function 3.1.2.1. But here we will use the result of a solver run to choose the constraints to add to the next run.

A birds-eye view: We execute a model with no subtour elimination constraints. If the solver returns a tour, we are done. If it returns a set of subtours, we add subtour elimination constraints for each of them, and only them. Eventually all *relevant* subtours are eliminated and the solver returns a tour of the whole graph. It takes longer to explain this approach than to implement it by writing a few lines of code.

5.4.1.4 Executable Model

Let's translate this into executable code which we will split into two: a first model which, given some set of subtours, will optimize after adding subtour elimination constraints for that particular set. See Listing 5-5. A second one, a main routine, iteratively calls the first model, adding subtours as they appear.

Listing 5-5. TSP Model with Subtour Elimination Constraints (tsp.py)

```
1  def solve_model_eliminate(D,Subtours=[]):
2    s,n = newSolver('TSP', True),len(D)
3    x = [[s.IntVar(0,0 if D[i][j] is None else 1,")) \
4          for j in range(n)] for i in range(n)]
5    for i in range(n):
```

```
6       s.Add(1 == sum(x[i][j] for j in range(n)))
7       s.Add(1 == sum(x[j][i] for j in range(n)))
8       s.Add(0 == x[i][i])
9     for sub in Subtours:
10      K = [x[sub[i]][sub[j]]+x[sub[j]][sub[i]]\
11          for i in range(len(sub)-1) for j in
            range(i+1,len(sub))]
12      s.Add(len(sub)-1 >= sum(K))
13    s.Minimize(s.Sum(x[i][j]*(0 if D[i][j] is None else D[i][j]) \
14                for i in range(n) for j in range(n)))
15    rc = s.Solve()
16    tours = extract_tours(SolVal(x),n)
17    return rc,ObjVal(s),tours
```

Line 4 defines the decision variable, a binary indicator of the arcs to take. The loop starting at line 5 enforces that there must be an arc in and an arc out of every node, exactly as in equations (5.6)-(5.7). We also enforce that all x[i][i] are zero to avoid loops. For each subtour provided by the caller, we extract all arcs of the corresponding clique at line 9 and constrain the sum of them to be one less than the number of vertices in the clique. We process the solution returned by the solver to extract, at line 16, the subtours and return them to the caller. This extraction code is shown in Listing 5-6.

Listing 5-6. Subtour Extraction

```
1  def extract_tours(R,n):
2    node,tours,allnodes = 0,[[0]],[0]+[1]*(n-1)
3    while sum(allnodes) > 0:
4      next = [i for i in range(n) if R[node][i]==1][0]
5      if next not in tours[-1]:
6        tours[-1].append(next)
7        node = next
```

```
8      else:
9          node = allnodes.index(1)
10         tours.append([node])
11     allnodes[node] = 0
12   return tours
```

The main loop is simple: we iterate until the number of tours returned by the solver is one, taking care to accumulate subtours as they are discovered. See Listing 5-7.

Listing 5-7. TSP Model Mainline (`tsp.py`)

```
1  def solve_model(D):
2    subtours,tours = [],[]
3    while len(tours) != 1:
4      rc,Value,tours=solve_model_eliminate(D,subtours)
5      if rc == 0:
6        subtours.extend(tours)
7    return rc,Value,tours[0]
```

The solution to our small example, in Table 5-10, one iteration per row, illustrates the subtours that were eliminated. In parentheses, we see the optimal value, the total length. As we eliminate subtours, it increases, of course. A graphical representation is shown in Figure 5-1 where the subtours eliminated at each iteration are shown in distinct shades.

Table 5-10. *Successive Iterations of the TSP Solver Showing Optimal Values and Subtours*

Iter (value)	Tour(s)
0-(2177)	[0, 2]; [1, 7, 8]; [3, 6, 9]; [4, 5]
1-(2526)	[0, 2, 7]; [1, 8]; [3, 6]; [4, 9, 5]
2-(2673)	[0, 2, 3, 6, 9, 5, 4, 1, 8, 7]

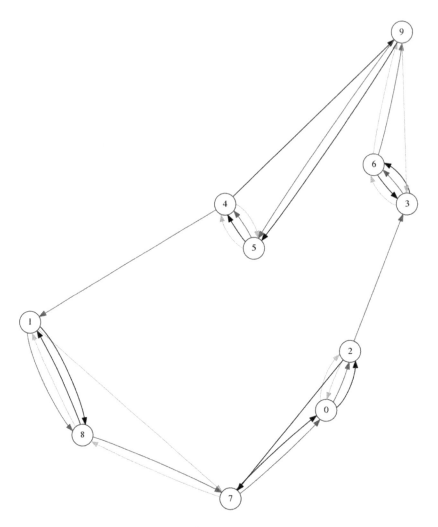

Figure 5-1. *Successive (partial) solutions of TSP example*

The key idea to take away from the TSP model is not that we can solve gigantic instances with it (specialized algorithms will do much better).[9] It is that, even if the number of constraints to completely specify a correct model is large, it may be possible to include a very small fraction of the

---

[9]See www.math.uwaterloo.ca/tsp/concorde.html for instance

required constraints and yet solve the problem to optimality. To do this effectively, one must have a deep understanding of the problem and good practical modeling skills (and of course, a good modeling language and library, such as Python and OR-Tools).

As a final note, the reader should be aware that our subtour elimination constraints are not the only ones possible. There are a number of ways to eliminate subtours, but none are as simple to implement as those we described. In practice, problems that have a TSP subproblem embedded in them have a number of other requirements, making the model relatively complex. Having a simple yet effective way to deal with subtours is a skill required of all modelers.

5.4.2 Variations

The TSP is one of the most studied combinatorial problems, hence it has a number of variations.

- A simple variation occurring often is that, instead of a tour (a closed path), one wants a simple path covering all vertices. For reference, let's call this problem TSP-P. Since we know how to solve the tour problem, the easiest way to solve the path problem is to transform the latter into the former. We add another node to the network; let's call it the dummy node. We also add arcs of distance zero from the dummy node to every other node on the network. Then we solve the TSP on that new network. At optimality we will have a tour going into and out of the dummy node. Deleting those two arcs yields the required path. Code to implement this variation is shown in Listing 5-8 and the result of running the path model on our example yields Table 5-11.

Listing 5-8. Code to Solve the TSP-P Problem (`tsp.py`)

```
1  def solve_model_p(D):
2    n,n1 = len(D),len(D)+1
3    E = [[0 if n in (i,j) else D[i][j] \
4      for j in range(n1)] for i in range(n1)]
5    rc,Value,tour = solve_model(E)
6    i = tour.index(n)
7    path = [tour[j] for j in range(i+1,n1)]+\
8           [tour[j] for j in range(i)]
9    return rc,Value,path
```

Table 5-11. *Result of the TSP-P Path Model on Our Example*

Nodes	1	8	7	0	2	5	4	6	3	9
Distance	0	295	350	221	107	264	93	268	101	302
Cumulative	0	295	645	866	973	1237	1330	1598	1699	2001

- A more complex variation is to allow repeated visits to nodes. For reference, let's call this problem as TSP*. The justification for this problem is simple: it is conceivable that one could find a shorter overall walk if one is allowed to visit any node more than once.

  Again the trick is to rely on our TSP model by transforming TSP* into TSP on a different network. The new network has exactly the same nodes, but the distance between the nodes is that of a shortest path between nodes of the original network.

We must take care to keep track of these shortest paths to reconstruct the TSP* solution. Since we already implemented an all-pairs shortest paths model (Listing 4-7 in Chapter 4), we will use it here. Code to implement TSP* is shown in Listing 5-9 and its solution on our example is shown in Table 5-12 and in Figure 5-2. Note that the total length is less than the TSP length even though it repeats some nodes.

Listing 5-9. Code to Solve the TSP* Problem (`tsp.py`)

```
1   def solve_model_star(D):
2       import shortest_path
3       n = len(D)
4       Paths, Costs = shortest_path.solve_all_pairs(D)
5       rc,Value,tour = solve_model(Costs)
6       Tour=[]
7       for i in range(len(tour)):
8         Tour.extend(Paths[tour[i]][tour[(i+1) % len(tour)]][0:-1])
9       return rc,Value,Tour
```

Table 5-12. *Result of the TSP* Function on Our Example*

NB 12		0	2	3	6	9	5	4	5	1	8	7	0
Total dist 2636	0	107	281	93	238	434	93	118	406	295	350	221	

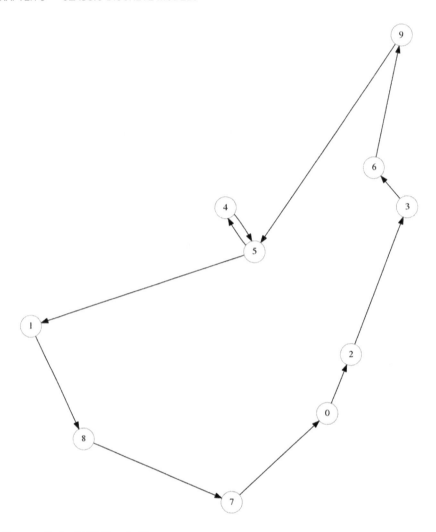

Figure 5-2. *TSP* solution*

CHAPTER 6

Classic Mixed Models

This section we will develop models for problems requiring a mix of continuous and integral variables, as well as a variety of constraints. Traditionally these models have been called mixed integer programs (MIP).[1]

I will cover a few of the classical problems that require a combination of continuous and discrete variables. The prototypical and simplest situation, exhibited by *facility location*, is when we have continuous objects floating around a network, but where the existence of the network nodes is subject to an optimizing decision. At the other extreme, one of the hardest situations is the *job shop* scheduling problem, where we want to construct a sequence of operations of various machines subject to precedence constraints.

These mixed models vary greatly in difficulty. There is practically no limit to the size of problems we can solve in the former instance, yet problems of the latter type with a dozen elements will prove, in practice, almost intractable.

6.1 Facility Location

Let's revisit the distribution problems we first encountered while discussing flows, but with a few added twists. Recall Section 4.2 of Chapter 4 where Solar-1138 Inc. needed to decide which plant would distribute power to

---

[1]Another questionable moniker. The variables are mixed, as some are continuous and other integral. Logic suggests that the name be mixed program (MP) or mixed continuous integer program (MCIP).

© Serge Kruk 2018
S. Kruk, *Practical Python AI Projects*, https://doi.org/10.1007/978-1-4842-3423-5_6

which city. Now let's assume, in addition, that Solar-1138 is in the planning stages and needs to decide first which plants to build to then distribute power to cities. The data is similar. First is the matrix of cost from each potential plant to each city, as shown in Table 6-1.

Table 6-1. *Example of Distribution Cost*

From/To	City0	City1	City2	City3	City4	City5	City6	Supply
Plant 0	20	23	23	24	28	25	13	544
Plant 1	19	18	30	25	19	17	14	621
Plant 2	29	13	19	17	22	15	11	635
Plant 3	16	23	29	22	29	26	11	549
Plant 4	23	20	10	27	23	19	20	534
Plant 5	21	12	23	29	14	15	22	676
Plant 6	13	18	22	13	11	25	23	616
Plant 7	21	12	20	20	20	13	11	603
Plant 8	24	24	29	17	18	16	20	634
Plant 9	28	11	22	26	25	19	11	564
Demand	553	592	472	495	504	437	634	

In addition, we have a cost to build each potential plant, in shown in Table 6-2. Since the cost of establishing the plants, as well as the cost of distribution, varies with the plants, we have a more complex problem on our hands than simple distribution. It's more complex both in terms of model and of solution technique. The questions to answer now are which plants to build and how much power to send from each plant to each city. I will forego any discussion of amortization, assuming that the appropriate calculations led to the fixed costs.

Table 6-2. *Example of Plant Building Costs*

Plant	0	1	2	3	4	5	6	7	8	9
Cost	5009	5215	6430	5998	4832	6365	6099	5499	5217	6153

6.1.1 Constructing a Model

6.1.1.1 Decision Variables

Since we have two related but distinct decisions, we need two sets of decision variables. As with the previous distribution model, we will need to know how much from plant i goes to city j, therefore

$$x_{i,j} \ \forall i \in P, \forall j \in C$$

As usual with distribution, the interpretation of, say, $x_{2,4} = 5$ will be to send 5 units from plant 2 to city 4. In our case, this is a continuous variable. We consider it appropriate to send fractional values of power across the network. For some applications, it will be an integer variable.

Since we also need to know if plant i will be built, we need a binary variable,

$$y_i \in \{0,1\} \ \forall i \in P,$$

with the interpretation that $y_2 = 1$ will indicate that plant 2 must be built.

6.1.1.2 Objective

The objective now has two parts, traditionally known as the fixed and variable costs. The fixed costs are those related to the plant building. The variable costs are the distribution costs, as in equation (4.4), to which we add the fixed cost. Plant i will be built only if variable y_i is one, assuming that the costs are in c_i,

$$\min \sum_i \sum_j C_{i,j} x_{i,j} + \sum_i c_i y_i$$

6.1.1.3 Constraints

As with the traditional minimum cost flow problem, we must have the supply and demand constraints,

$$\sum_j x_{i,j} \le S_i \quad \forall i \in P, \tag{6.1}$$

and

$$\sum_i x_{i,j} = D_j \quad \forall j \in C \tag{6.2}$$

Now we need to consider the new element of this model. How do we link the variables related to plant building to those related to power distribution? We must not distribute from a plant that is not built and we must not build plants from which we will not distribute anything.

Since the objective function minimizes, it will tend to set all variables to zero, unless it cannot (assuming, of course, that costs are positive). We already know, from our work in distribution, that the $x_{i,j}$ variables will be properly set. Therefore, we must ensure that the corresponding y_i variables are also set.

When should a particular y_i be one, indicating that plant i is to be built? When the optimal solution has some $x_{i,j}$ above zero for the same i and any j. This suggests a constraint of the form

$$\sum_j x_{i,j} \le y_i \quad \forall i \in P$$

This constraint achieves half of our needs: no $x_{i,j}$ will be above zero if plant i is not built. Yet, as y_i is at most one, while the sum on the left side may be considerably larger, there is an inconsistency. The trick is to multiply y_i by a "sufficiently" large constant, M. What would be large enough? The sum of the demands of all cities would certainly work since the sum of all the $x_{i,j}$ cannot possibly exceed all the demands. Therefore, assuming that the sum of all demands is M,

$$\sum_j x_{i,j} \leq My_i \ \forall i \in P \tag{6.3}$$

Constraints of type (6.3) are known to optimizers as big-M constraints[2] and are to be used sparingly unless constants M are small enough. Solvers, some more than others, may get into numerical trouble if the constants are too large. In practice, this means the modeler should find the smallest possible M and try it. If the solver chokes, then find another modelling technique.

A different big-M approach may occur to the reader: since any of the $x_{i,j}$ of non-zero should trigger the corresponding y_i to be one, this set of constraints is possible:

$$x_{i,j} \leq Dy_i \ \forall i \in P \ \forall j \in C$$

for some suitable multiplier D. Indeed, this solves our problem and considerably reduces the size of the multiplier at the cost of increasing the number of constraints.[3]

The model as described above will work, but we can improve it both by reducing M and by eliminating some constraints. Notice that constraints (6.1) and (6.3) have the same structure and the same left-hand side. This condition suggests a merge of the constraints into

$$\sum_j x_{i,j} \leq S_i y_i \ \forall i \in P \tag{6.4}$$

Indeed, had we given the size of M more thought, we might have concluded that Si was the "best" big-M to use.

---

[2] Yes, more sadly unimaginative nomenclature.
[3] For the theoretically-minded, there are cases where multiplying the number of constraints is actually preferable as it provides a tighter relaxation; this is not one of those cases.

6.1.1.4 Executable Model

The reader will recognize most of this model as it is identical to Listing 4-2. I will only highlight the differences.

The solver receives, in addition to the distribution cost matrix *D*, which includes the demand and supply data, an array of fixed building costs, *F*.

Listing 6-1. Facility Location Model (`facility_location.py`)

```
1  def solve_model(D,F):
2    s = newSolver('Facility_location_problem', True)
3    m,n = len(D)-1,len(D[0])-1
4    B = sum(D[-1][j]*max(D[i][j] \
5      for i in range(m)) for j in range(n))
6    x = [[s.NumVar(0,D[i][-1],") for j in range(n)] \
7        for i in range(m)]
8    y = [s.IntVar(0,1,") for i in range(m)]
9    Fcost, Dcost = s.NumVar(0,B,"),s.NumVar(0,B,")
10   for i in range(m):
11     s.Add(D[i][-1]*y[i] >= sum(x[i][j] for j in range(n)))
12   for j in range(n):
13     s.Add(D[-1][j] == sum(x[i][j] for i in range(m)))
14   s.Add(sum(y[i]*F[i] for i in range(m)) == Fcost)
15   s.Add(sum(x[i][j]*D[i][j] \
16       for i in range(m) for j in range(n)) == Dcost)
17   s.Minimize(Dcost + Fcost)
18   rc  = s.Solve()
19   return rc,ObjVal(s),SolVal(x),SolVal(y),\
20          SolVal(Fcost),SolVal(Dcost)
```

Line 11 links the decision to build a given plant to the amount transported from that plant. Note that if *y[i]* is zero, indicating that we do not build plant *i*, then the corresponding *x[i][..]* will all be zero, so that no product will flow out of that plant. In the other direction, if $y[i]$ is one, then the product flowing out of the plant will never exceed its supply capacity $D[i][-1]$.

The objective function at line 17 minimizes sum of the fixed building costs and the variable distribution costs.

Finally, we return all the transported material as well as the decisions to build. The solution to this example appears in Table 6-3 where it displays only the transport from the plants included in the building decision.

Table 6-3. *Optimal Solution to the Facility Location*

	City 0	City 1	City 2	City 3	City 4	City 5	City 6
Plant 2		1.0					634.0
Plant 4			472.0			51.0	
Plant 5		235.0			441.0		
Plant 6	553.0				63.0		
Plant 7		356.0				247.0	
Plant 8				495.0		139.0	

6.1.2 Variations

- The main variation has to do with capacities. It may be that the path between producers and consumers has a maximum capacity, say $c_{i,j}$. This is trivially implemented by defining the variables with the appropriate domain, that is $0 \leq x_{i,j} \leq ci,j$ or, in the executable, amend line 7 to read

```
x = [[s.NumVar(0,C[i][j],") for j in range(n)]
    for i in range(m)]
```

In the case of a network with capacities, it may be worthwhile to revisit the big-M constraint used to set the building decision variable and prefer an alternative approach based on the capacity of the flow out of the plants.

6.2 Multi-Commodity Flow

Previously discussed flow problems were easy integer problems because the integrality came for free. There was no need to declare variables as integers to get an integral optimal solution. But this is not the case when multiple *commodities* are carried by the same network; then we must explicitly specify all variables that must be integral.

We can think of this problem as a series of transshipment problems overlaid on one network. Some nodes supply, some nodes demand, others can carry through, and there is more than one element to carry so that a node acting as a supplier for one element can be a consumer for another. As such, we will have a number of cost, demand, and supply data tables, as shown in Table 6-4 for a small instance. The goal is, as in transshipment, to satisfy all demands.

Table 6-4. *Example of Multi-Commodity Flow Cost Matrices*

Comm 0	N0	N1	N2	N3	N4	Supply
N0		20	23	23	24	532
N1	19		18	30	25	
N2				13	19	
N3	24	23			22	512
N4	23	10				
Demand		230	306		508	

(continued)

Table 6-4. (*continued*)

Comm 1	N0	N1	N2	N3	N4	Supply
N0		23	29	14	15	533
N1	22		13	11	25	609
N2	20	20		13	11	634
N3		18	20		24	
N4			11	22		
Demand				354	1422	
Comm 2	**N0**	**N1**	**N2**	**N3**	**N4**	**Supply**
N0		30	17	19	30	
N1			21	19	27	564
N2				29	27	588
N3			12		27	
N4	27	15		16		
Demand	315			360	477	

6.2.1 Constructing a Model

The model will be described in stages.

6.2.1.1 Decision Variables

As it simply is a set of transshipment problems on the same network, we need a decision variable per problem. So, if we assume K commodities, and a set of N nodes,

$$x_{k,i,j} \in [0,1,2,\ldots] \quad \forall i \in N, \forall j \in N, \forall k \in K$$

If variable $x_{4,3,5}$ is 6, it will mean to send 6 units of commodity 4 along arc (3, 5).

6.2.1.2 Objective

The objective is a simple generalization of the transshipment objective,

$$\min \sum_k \sum_i \sum_j C_{k,i,j} x_{k,i,j}$$

where we now minimize the cost of shipping all commodities across the network.

6.2.1.3 Constraints

The constraints are the generalized conservation of flow, taking care that they must all hold for each commodity separately:

$$\sum_j x_{k,j,i} - \sum_j x_{k,i,j} = D_{k,i} - S_{k,i} \quad \forall k \in K, i \in N \tag{6.5}$$

6.2.1.4 Executable Model

Let's translate this into an executable model, shown in Listing 6-2. The function accepts a three-dimensional array C of costs on arcs for each commodity. It also accepts capacities in D that can be either a scalar, indicating the same capacity on all arcs, or a two-dimensional array to specify a capacity per arc. Finally, it accepts parameter Z to indicate, if True, that we must solve this as an integer program because the elements transported by the network are indivisible. If False, we are accepting fractional solutions. The reason for this last parameter is that accepting fractional solutions, while it accelerates the process tremendously, may still produce integral solutions.

Listing 6-2. Multi-Commodity Flow Model (multi_commodityflow.py)

```
1  def solve_model(C,D=None,Z=False):
2      s = newSolver('Multi-commodityumincostuflowuproblem', Z)
3      K,n = len(C),len(C[0])-1,
```

```
4     B = [sum(C[k][-1][j] for j in range(n)) for k in range(K)]
5     x = [[[s.IntVar(0,B[k] if C[k][i][j] else 0,") \
6           if Z else s.NumVar(0,B[k] if C[k][i][j] else 0,") \
7           for j in range(n)] for i in range(n)] for k in
            range(K)]
8     for k in range(K):
9       for i in range(n):
10        s.Add(C[k][i][-1] - C[k][-1][i] ==
11              sum(x[k][i][j] for j in range(n)) - \
12              sum(x[k][j][i] for j in range(n)))
13    if D:
14      for i in range(n):
15        for j in range(n):
16          s.Add(sum(x[k][i][j] for k in range(K)) <= \
17              D  if type(D) in [int,float] else D[i][j])
18    Cost = s.Sum(C[k][i][j]*x[k][i][j] if C[k][i][j] else 0\
19          for i in range(n) for j in range(n) for k in range(K))
20    s.Minimize(Cost)
21    rc  = s.Solve()
22    return  rc,ObjVal(s),SolVal(x)
```

This code is essentially identical to the transshipment code, with some practical embellishments. The decision variable now has three dimensions instead of two: the first one indicating which commodity, and the last two, as usual, the arc. Moreover, if the parameter Z is True, it is defined as an integer variable. The reason for the choice is that, for a very large number of networks (the vast majority, in fact) and multicommodity flow problems, the variables can be declared as continuous and yet, if all demands, supplies, and capacity are integral, so will the solution be. The modeler who is faced with long runtimes should try to relax the integrality constraint. It might very well be integral, saving oodles of CPU cycles. The conditions under which solutions will be integral are complex and difficult to verify ahead of time, which is why, for practical purposes, it is easier to try a continuous solver and adjust if the solution is not practical.

The solution to this simple problem is shown in Table 6-5. And, in fact, it was solved as a continuous problem.

Table 6-5. *Optimal Solution to the Multi-Commodity Flow*

Comm 0	N0	N1	N2	N3	N4
N0		226	306		
N1					
N2					
N3		4			508
N4					
Comm 1	N0	N1	N2	N3	N4
N0					533
N1			155	354	100
N2					789
N3					
N4					
Comm 2	N0	N1	N2	N3	N4
N0					
N1				360	204
N2					588
N3					
N4	315				

6.2.2 Variations

6.2.2.1 All-Pairs Shortest Paths (Revisited)

Why did I cover multi-commodity flows if they are such trivial modifications to transshipment? Because they are often used by twisting a problem to fit the structure of multi-commodity flows.[4] Here is a simple example: remember the problem of finding the shortest paths between all pairs of nodes in a network? We did it by solving a sequence of the shortest paths model. It is possible to find all such paths by running a single multicommodity flow; better yet, it will **never** require an integrality constraint, ensuring a very short runtime.

The trick is to consider every node to be a supplier of a particular commodity in quantity $n - 1$. And every node is also a consumer of the $n - 1$ commodities provided by the other nodes. Listing 6-3 implements this approach. The code is slightly more general than that because it allows us to specify a set of source nodes for which we want emanating paths to all other nodes.

Running the code on the identical problem to the all-pairs problem in Chapter 4 yields identical results. (See Table 4-11). The difference is that the flow code is much faster. It is, in fact, often much faster than running some specialized algorithm like Floyd-Warshall, which has a fixed complexity proportional to $n3$, while our flow problem is often solved in time proportional to n.

Listing 6-3. All-Pairs Shortest Paths via Multi-Commodity Flow

```
1  def solve_all_pairs(D,sources=None):
2    n,C = len(D),[]
3    if sources is None:
4      sources = [i for i in range(n)]
```

---

[4]There are a number of research papers on creating timetables using multi-commodity flows, for instance.

```
5    for node in sources:
6      C0 = [[0 if n in [i,j] else D[i][j] for j in range(n+1) ] \
7              for i in range(n+1)]
8      C0[node][-1] = n-1
9      for j in range(n):
10      if j!= node:
11         C0[-1][j] = 1
12       C.append(C0)
13     rc,Val,x = solve_model(C)
14     Paths = [[None for _ in range(n)] for _ in sources]
15     Costs = [[0 for _ in range(n)] for _ in sources]
16     if rc == 0:
17       for source in sources:
18        ix = sources.index(source)
19        for target in range(n):
20         if source != target:
21            Path,Cost,node = [target],0,target
22            while  Path[0] != source  and  len(Path)<n:
23              v = [j for j in range(n) if x[ix][j][node]>=0.1][0]
24              Path.insert(0,v)
25              Cost  +=  D[v][node]
26              node = v
27           Paths[ix][target] = Path
28           Costs[ix][target] = Cost
29     return rc, Paths, Costs
```

Table 6-6 shows the results of running the code and requesting the shortest paths emanating from nodes 0 and 2. The reader should see that the lengths are identical to those of Table 4-12.

Table 6-6. *Shortest Paths from Nodes 0 and Node 2 in the Example of Section 4.4*

0-Target	Cost	[Path]
1	46	[0, 1]
2	17	[0, 2]
3	24	[0, 3]
4	51	[0, 4]
5	48	[0, 2, 5]
6	52	[0, 2, 5, 6]
7	41	[0, 3, 7]
8	68	[0, 2, 8]
9	55	[0, 3, 7, 9]
10	83	[0, 3, 7, 9, 10]
11	99	[0, 2, 5, 11]
12	89	[0, 3, 7, 9, 10, 12]
2-Target	**Cost**	**[Path]**
0	79	[2, 5, 0]
1	38	[2, 1]
3	68	[2, 5, 6, 3]
4	34	[2, 4]
5	31	[2, 5]
6	35	[2, 5, 6]
7	58	[2, 5, 7]

(*continued*)

Table 6-6. (*continued*)

2-Target	Cost	[Path]
8	51	[2, 8]
9	66	[2, 5, 9]
10	88	[2, 5, 10]
11	82	[2, 5, 11]
12	94	[2, 5, 10, 12]

6.2.3 Instances

An interesting application appears in fiber optics networks. Consider a set of sources where signals emanate and a set of sinks that these signals must reach and a set of transshipment nodes that only carry signals (possibly boosting them if needed). It is possible for multiple signals to share the same cables at the same time if they use different wavelengths. The number of available wavelengths is limited. What we have here is multiple transshipment problems overlaid. The objective is, in this case, to maximize the number of established connections.

6.3 Staffing Level

This problem is often described by optimizers as a *staff scheduling problem*. But that is a terrible misnomer as no schedule is ever produced, only requirement levels. The true staff scheduling problem as it is known to practitioners is orders of magnitude more complex. So we will call this problem the *staffing level* problem.

The situation is the following: we have a grid, indexed in one dimension by time intervals. This grid is most commonly presented in either days (Monday, Tuesday) or hours (8AM, 9AM) but could be in any valid time units. In the other dimension, we have what are usually called shifts, units of a worker's schedule (Monday to Friday, Tuesday to Sunday, or 8AM to 4PM, 9AM to 5PM). Associated with each time interval we have staffing needs (need 45 people on Monday or 62 people at noon). Finally, we also have a cost associated to each shift. See Table 6-7 for an example.

Table 6-7. *Staffing Requirement Matrix*

	Shift 0	Shift 1	Shift 2	Shift 3	Shift 4	Shift 5	Shift 6	Need
00h	1			1				15
02h	1					1		17
04h	1					1		16
06h	1	1						17
08h		1					1	19
10h		1					1	18
12h		1	1					11
14h			1					12
16h			1					15
18h			1	1				14
20h				1	1			20
22h				1	1			18
Cost	$69.37	$67.03	$64.55	$72.06	$29.24	$21.67	$24.52	

Note that the example includes two types of shifts: full-time (eight hours) and part time (four hours). The costs and the number of hours per shift are different because typically full-time workers are better paid than part-timers. The goal is to determine, at the least total cost, how many people will work each shift, ensuring that the needs are satisfied and also that the full-time workers get preferential treatment. This preference can take on various forms. It can be as simple as "No part-timer if there is no full-timer working" or "Every full-time shift must be staffed by at least x people" or "We have a pool of x full-time staff who must work; fill the rest of the needs with part-timers." We will discuss some possible variations on these constraints.

Note that a solution to this problem determines only the staffing level of each shift. This is not a schedule. The true schedule, in the sense of which worker works which shift, is left to a further model, which is a much more complicated one.

The reader should notice here a common approach of optimizers for complex problems: decomposition. The staffing needs shown in the last column of Table 6-7 were conceivably obtained by some model. The levels will be produced by the model we are now creating. Finally, the proper schedules will be generated by yet another model. This approach is unlikely to produce the overall optimal solution. Yet it is used in all industries. The airlines are notorious for decomposing problems. There are two reasons we approach complex problems this way: a good reason and a bad one.

The good reason is that, often, attacking the whole problem is not technologically feasible. We can write the whole problem model, certainly, but no solver can find a solution before the solar system expires. Another aspect of the same reason is that one part of the whole problem is best modeled as an integer program while another is much simpler as a constraint program. There are yet no entirely satisfying ways to write such hybrid models.[5] The second reason, the bad one, is that, even in cases where a more holistic approach is feasible, there is massive inertia in organizations that would prevent its implementation.[6]

---

[5]It is being done, piecemeal, by very talented modelers willing to dig deep into the bowels of solvers but there is no easy way to do this in general yet. Ask me again in five years; OR-Tools is very close to that goal.

[6]The paper pushers responsible for one aspect of the problem will fight tooth and nail the paper pushers responsible for another aspect, lest they relinquish part of their fiefdom. I have seen massive savings of time and money squandered in internecine turf wars.

6.3.1 Constructing a Model

The model will be described in stages.

6.3.1.1 Decision Variables

What we need to decide in this problem is how many people need to work each shift. Because of the distinction between full-time and part-time shifts, it may be useful to distinguish between an employee working full time (FTE) and an employee working part time (PTE). Hence, assuming N_f and N_p for the sets of full-time and part-time shifts,

$$x_i \in [0,1,2,\ldots] \; \forall i \in N_f \cup N_p =: N$$

6.3.1.2 Objective

The objective is simple since we have a cost per shift C_i,

$$\min \sum_{i \in N} C_i x_i$$

6.3.1.3 Constraints

Let's tackle the staffing needs by looking at a specific time interval, say 0600h with a need of 17 people. If shift 0 covers from midnight to 8AM, and shift 1 covers 6AM to 2PM, both of those shifts, and no other, must together have at least 17 people to cover the 6AM need. Algebraically, it's

$$x_0 + x_1 \geq 17$$

Now, in more general terms, consider a matrix $M_{T \times N}$ where T is the set of time intervals, with the structure of Table 6-7 (without the last row and column). What we need is for every time interval t, the sum of the x_i that covers t is at least the required R_t. Algebraically, it's

$$\sum_{i \in N} M_{t,i} x_i \geq R_t \;\; \forall t \in T \tag{6.6}$$

This is enough for the simplest staffing level model. If there is no distinction between full and part time, we are done. Let's consider a few realistic constraints.

1. *Minimum full-time staff:*

   If we have a requirement that, for every full-time shift i, there be at least Q_i people working, we can add

   $$x_i \geq Q_i \;\; \forall i \in N_f$$

   Because of the structure of the shifts (full-time shifts cover all time intervals), this constraint also satisfies a requirement of the form "No PTE working if no FTE are working" (assuming that $Qi > 0$, of course).

2. *FTE must work:*

   A possibly more realistic requirement is that, given a pool of P of full-time employees (FTE), we require that they all work. Only if the staffing requirements cannot be met should we involve part-time staff. This is accomplished by

   $$\sum_{i \in N_f} x_i = P \tag{6.7}$$

   This constraint will be infeasible unless the number of full-time workers is below or at the staffing needs, of course.

3. *No PTE if no FTE present:*

Consider a particular time interval, say 10AM to noon. Say it can be staffed by full-time shift 1 and 2 and also by part-time shift 6. We want to ensure that x_6 is zero unless either x_1 or x_2 is non-zero. As a first attempt, let's try

$$x_1 + x_2 \geq x_6 \qquad (6.8)$$

This ensures that if there are no full-time staff in either full-time shift, then there will be no part-time staff either. But it constrains the problem more than necessary because it prevents the number of part-time staff to exceed the full-time staff. What we need is to scale up the left-hand side of equation (6.8) by some "large enough" constant.

This is again an instance of the technique optimizers call the big-M method. How large is large enough? The sum of the required staff is surely enough and yet not so large as to cause numerical problem. Therefore,

$$\sum_{j \in T} R_j \sum_{i \in N_f} M_{t,i} x_i \geq \sum_{i \in N_p} M_{t,i} x_i \ \forall t \in T \qquad (6.9)$$

For each time interval, equation (6.8) sums, on the left, the number of full-time workers, scaled up by a large constant. On the right it sums the number of part-time workers.

6.3.1.4 Executable Model

The model will receive a time vs. shift 0-1 matrix M, exactly as in Table 6-4, along with the number of shifts considered full-time. This number will be equal to the number of columns of M if the problem has no part-time staff. The last column indicates minimal requirements for each time interval. The last row indicates cost of one worker in the indicated shift.

To make this a more practical model, it may be given an array Q, indexed by shifts indicating the minimum number of people in each shift. It may also be given a scalar P of full-time people who must work. Finally, it may also be given a flag no part which, if True, indicates that there will be no part-time people working unless full-time people are present.

Listing 6-4. Staffing model (staffing.py)

```
1   def solve_model(M,nf,Q=None,P=None,no_part=False):
2       s = newSolver('Staffing', True)
3       nbt,n = len(M)-1,len(M[0])-1
4       B = sum(M[t][-1] for t in range(len(M)-1))
5       x = [s.IntVar(0,B,'') for i in range(n)]
6       for t in range(nbt):
7           s.Add(sum([M[t][i] * x[i] for i in range(n)]) >= M[t][-1])
8       if Q:
9           for i in range(n):
10              s.Add(x[i]  >=  Q[i])
11      if P:
12          s.Add(sum(x[i] for i in range(nf)) >= P)
13      if no_part:
14          for t in range(nbt):
15              s.Add(B*sum([M[t][i] * x[i] for i in range(nf)]) \
16                      >= sum([M[t][i] * x[i] for i in range(nf,n)]))
```

```
17    s.Minimize(sum(x[i]*M[-1][i] for i in range(n)))
18    rc  = s.Solve()
19    return rc, ObjVal(s), SolVal(x)
```

After extracting the number of time intervals, number of shifts, and computing a bound for the decision variables on line 4, we define our decision variable as a non-negative integer. The bound we use is clearly valid as it is the sum of all required staff.

We implement the basic constraint of equation (6.6) at line 7. Then, the three following if statements implement the optional constraints of *Minimum full-time staff*, *FTE must work*, and *No PTE if no FTE present*.

Results of the basic constraint only are shown in Table 6-8 while setting the no part flag to True yields Table 6-9. Note that, for this example, the total number of people working has not changed, but the distribution of people among shifts has changed, raising the total cost.

Table 6-8. *Optimal Solution to the Basic Staffing Level Problem*

$3187.84	Shift 0	Shift 1	Shift 2	Shift 3	Shift 4	Shift 5	Shift 6
Nb:71	15	2	15	0	20	2	17

Table 6-9. *Optimal Solution with "No Part-Time Unless Full-Time" Constraint*

$3225.47	Shift 0	Shift 1	Shift 2	Shift 3	Shift 4	Shift 5	Shift 6
Nb:71	14	3	15	1	19	3	16

6.3.2 Variations

We have covered some variations by implementing optional parameters already. There are as many other variations as there are companies using staffing levels. The next step involves transforming this into a real scheduling model, assigning specific people to shifts. We will begin to tackle this later, but the reader is encouraged to ponder ways to implement true scheduling.

6.4 Job Shop Scheduling

A rather difficult problem to solve is the following: consider set of jobs *J* to be performed on set of machines *M*. Each job requires some time on each of the machines to perform a specific task and has an order in which the tasks must be performed. You can think of building wooden toys. The wood needs to be cut to shape, then sanded, then primed, then painted, then covered in lacquer. Each of these tasks is accomplished by a different machine and requires a certain duration, which depends on the toy. The word "machine" here is used in a rather wide sense. It could indicate a highly trained worker at his station. It could be a robot. The idea is the same. Moreover, not all jobs need all machines. Some may require only a subset of the machines. The overall goal is to schedule time on each machine to do all tasks while minimizing the overall time used.

The small example we will use to illustrate this model is given in Table 6-10.

Table 6-10. *Example of Job Shop Scheduling (for Each Job, the Machine and Duration of Tasks)*

Job	Machine-Duration	Machine-Duration	Machine-Duration
J0	1-10	2-10	0-10
J1	1-5	0-8	2-5
J2	2-5	1-9	0-9
J3	0-6	2-9	1-5

6.4.1 Constructing a Model

Since we are given the order of the tasks for each job, it is not part of the decision. The time at which a particular task is started for a given job on a given machine is what we seek. We will therefore use a decision variable indicating the starting time:

$$x_{ik} \in [0,\infty) \forall i \in J, \ \forall k \in M$$

We are not making the assumption that the durations are given as integers so that the decision variable is continuous.

The first type of constraints puts a lower bound on the starting times. Consider, say, that job 7 must use machine 4 for 3 hours before it goes to machine 6, then the starting time is $x_{74} + 3 \le x_{76}$. In general, assuming that p_i is the vector of order of machines of job i (a permutation of $0, 1, 2, \dots, M-1$) with duration vector d_i, we are led to

$$x_{ip_{ik}} + d_{ik} \le x_{ip_{ik+1}} \ \forall i \in J, \ \forall k \in \left[0,\dots,|M|-2\right]$$

The difficulty in this problem stems from enforcing that, at any given time, there is at most one job per machine. It is acceptable for a machine to be idle, though we will try to minimize idleness. How can we enforce this constraint?

One way is to introduce an additional variable that will indicate the relative order of jobs on a given machine. For instance,

$$z_{ijk} \in \{0,1\}, \ \forall i, j, \in J, \ \forall k \in M,$$

with the interpretation that $zijk = 1$ if and only if job i precedes job j on machine k. Note that, as either job i precedes job j or vice-versa,

$$z_{ijk} = 1 \Leftrightarrow z_{jik} = 0$$

With this variable we can try to enforce our difficult condition. Consider again a simple example. Say that job 7 needs machine 2 for 3 hours and job 5 needs it for 4 hours. Then either $x_{72} + 3 \leq x_{52}$, indicating that job 5 cannot start until job 7 has ended on machine 2, or else that $x_{52} + 4 \leq x_{72}$ indicating the reverse temporal condition. This is a disjunction; we need one or the other of the constraints to hold.

Using variables z_{ijk} this condition can be enforced by considering, for all jobs i, j and all machines k,

$$z_{ijk} = 1 - z_{jik}$$
$$x_{ip_{ik}} + d_{ik} - Mz_{ijp_{ik}} \leq x_{jp_{ik}}$$
$$x_{jp_{jk}} + d_{jk} - Mz_{jip_{jk}} \leq x_{ip_{jk}}$$

for some large enough value M. Note that only one of the two z_{ijk} or z_{ijk} will be one. Hence exactly one of the two inequalities will constraining. The other will be vacuously satisfied.

All that is left is to deal with the objective function. We want a schedule that will complete all jobs in the least amount of time. We could add a bound to the end of each job and minimize that bound. For instance,

$$x_{ip_{ik}} + d_{ip_{ik}} \leq T \forall i \in J, \forall k \in \left[0, \ldots, |M| - 1\right]$$

with objective

$$\min T$$

where T will be the completion time of the last task on the last machine.

6.4.1.1 Executable Model

The executable code is seen in Listing 6-5. It assumes that the input is a list of tuples exactly as in Table 6-10, indicating for each job the order of the machines needed with the duration of the task on each machine.

Listing 6-5. Job Shop Scheduling Model (`job shop.py`)

```
1   def solve_model(D):
2     s = newSolver('JobuShopuScheduling', True)
3     nJ,nM   =   len(D),len(D[0])
4     M = sum([D[i][k][1] for i in range(nJ) for k in
         range(nM)])
5     x = [[s.NumVar(0,M,'') for k in range(nM)] for i in
         range(nJ)]
6     p = [[D[i][k][0] for k in range(nM)] for i in range(nJ)]
7     d = [[D[i][k][1] for k in range(nM)] for i in range(nJ)]
8     z = [[[s.IntVar(0,1,'') for k in range(nM)] \
9        for j in range(nJ)] for i in range(nJ)]
10    T = s.NumVar(0,M,'')
11    for i in range(nJ):
12      for k in range(nM):
13        s.Add(x[i][p[i][k]] + d[i][k] <= T)
14        for j in range(nJ):
15          if i != j:
16            s.Add(z[i][j][k]   ==   1-z[j][i][k])
17            s.Add(x[i][p[i][k]] + d[i][k] - M*z[i][j][p[i][k]] \
18               <= x[j][p[i][k]])
```

```
19              s.Add(x[j][p[j][k]] + d[j][k] - M*z[j][i][p[j][k]] \
20                 <= x[i][p[j][k]])
21         for k in range(nM-1):
22             s.Add(x[i][p[i][k]] + d[i][k] <= x[i][p[i][k+1]])
23     s.Minimize(T)
24     rc = s.Solve()
25     return rc,SolVal(T),SolVal(x)
```

At line 4, we compute a value larger than the last possible end time by adding all durations of all tasks as if we had only one machine. This will be used to limit the domain of decision variables as well as bounding some constraints. Lines 6 and 7 extract from the input tuples a vector of permutation of the machines and a vector of durations. This will allow us to write the code as closely as possible to the abstract mathematical description we used.

The line 13 sets T as the upper bound on all the end times of all tasks. We will then minimize this T to obtain the shortest schedule. Line 22 is the translation of (6.4.1), enforcing that, for each job, the tasks are done in the order prescribed by the input tuples.

Finally, the if block starting at line 15 is where the complex disjunction occurs. It enforces that, for every pair of tasks, one strictly precedes the other. It is possible to reduce the number of variables by half, but a good solver will do that automatically during a pre-processing phase.

The result of our simple example is shown in Table 6-11 and Figure 6-1.

Table 6-11. *Optimal Solution (S is Start Time; M is Machine; D is Duration)*

	(S; M; D)	(S; M; D)	(S; M; D)
Job:0	{ 0.0; 1; 10 }	{ 13.0; 2; 10 }	{ 23.0; 0; 10 }
Job:1	{ 10.0; 1; 5 }	{ 15.0; 0; 8 }	{ 23.0; 2; 5 }
Job:2	{ 0.0; 2; 5 }	{ 24.0; 1; 9 }	{ 33.0; 0; 9 }
Job:3	{ 0.0; 0; 6 }	{ 28.0; 2; 9 }	{ 37.0; 1; 5 }

Time	0	1	2	3	4	5	6	7	8	9	10	1	2	3	4	5	6	7	8	9
Mach 0	3	3	3	3	3	3										1	1	1	1	1
Mach 1	0	0	0	0	0	0	0	0	0	0	1	1	1	1	1					
Mach 2	2	2	2	2	2	2								0	0	0	0	0	0	0

Time	20	1	2	3	4	5	6	7	8	9	30	1	2	3	4	5	6	7	8	9	40	1
Mach 0	1	1	1	0	0	0	0	0	0	0	0	0	0	2	2	2	2	2	2	2	2	2
Mach 1				2	2	2	2	2	2	2	2	2						3	3	3	3	3
Mach 2	0	0	0	1	1	1	1	3	3	3	3	3	3	3	3							

Figure 6-1. *Graphical representation of schedule*

CHAPTER 7

Advanced Techniques

In this chapter I cover some of the tricks optimizers have developed over the years to twist models and cajole solvers into providing solutions for problems that do not easily fit the Procrustean rules of mathematical optimization.

Some of these techniques involve iteratively solving a sequence of partial models, converging to the solution we seek. Some involve the creative use of layers upon layers of decision variables, each abstracting an additional level of details. Some involve the inspection of an invalid model to gain insight into an improved one. All in all, it's a mixed bag of jewels with little in common but the goal of solving more complex problems.

I end with a series of puzzles that are rarely considered by classical optimizers but are the bread and butter of constraint programmers. To solve them, we build a small library of functions that proves invaluable in succinctly expressing these models, proving, I trust, that with the right tools and the right language, the expressive power of constraint programming can be carried over to integer programming.

7.1 Cutting Stock

The cutting stock problem originated in the paper and textile industries. A typical mill produces paper in very large product rolls (large in length and in width), which are then cut to the width required by the customers. Let's call the latter *consumer rolls*. For instance, tabloid printers want their paper in rolls 17 inches wide, while broadsheet printers want twice that width. These consumer rolls are cut to page size after printing. The cutting problem, for the paper producer, is to cut the large-width rolls to satisfy customer demands while minimizing waste.

Table 7-1 is a random instance of orders we will solve to illustrate the model. The first column is the number of consumer rolls in the order and the second column is the required width that has been pre-processed to be measured in percentage of the product roll width.

Table 7-1. *Example of Cutting Stock Problem*

Order	Nb rolls	% Width
0	6	25
1	12	21
2	7	26
3	3	23
4	8	33
5	2	15
6	2	34

We want to minimize paper waste, which really means minimizing the total number of product rolls used. But, in order to do the cutting, we will need more information, namely given any product roll, how is it to be cut? Say we have a number of orders for 21% and for 36%, do we cut at 21%, 42%, 63%, and 99% with a waste of 1% or at 36%, 72%, and 93% for a waste of 7%? It may be that we need to cut half our rolls with the first pattern and half with the second to satisfy all demands. Which patterns to cut and how many rolls of a given pattern are the key problems. See Figure 7-1.

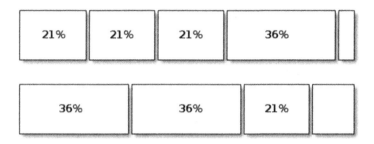

Figure 7-1. *Which pattern do we use?*

7.1.1 Constructing a Model

Originally the problem was attacked by specifying various cutting patterns, sometimes statically, sometimes dynamically. This was more or less forced by the technology of the time. Things have changed; both processors and solvers are considerably faster. Moreover, a good modeler should try the simplest approach first and make it more complex if, and only if, it fails. With this in mind, we will leave to the solver the problem of deciding on the patterns.

7.1.1.1 Decision Variables

If we leave the pattern decision to the solver, the problem becomes, given a roll, where do we cut? But this is over-specifying. The pattern 21, 42, 63, 99 is not distinguishable from the pattern 21, 57, 78, 99. They satisfy exactly the same customer demands: three 21% and one 36% width. And we know that having multiple indistinguishable solutions is very bad for a solver. Therefore, what we should ask is "Given a roll, how many cuts of width w do we make?"

With this in mind, assuming that we have N orders and at most K rolls, a reasonable decision variable is

$$x_{i,j} \in [0,1,2,\ldots] \, \forall i \in N, \forall j \in K,$$

where $x_{2,5} = 7$, for instance, will mean that we cut roll 5 to satisfy 7 customer rolls of width specified by order 2. The ordering of the cuts is irrelevant but we might very well post-process the solution to produce cutting patterns to be used.

And since we do not know ahead of time how many rolls we will need, there has to be a corresponding variable to indicate roll use,

$$y_j \in \{0,1\} \ \forall j \in K,$$

where $y_5 = 1$ means that roll 5 (out of a possible K) is used. This is the same trick we used in the facility location problem to decide which facility to open.

The maximum number of rolls, K, does not have to be precisely determined; any upper bound will do at this point.

7.1.1.2 Objective

The objective is to minimize the number of rolls. We could minimize the sum of all the y_j, but that leaves open the possibility of a solution that says "Use rolls 1 and 3, but not 2." To avoid such annoyance, let's use a small trick that will make each new roll more costly that the last:

$$\min \sum_{j \in K} j * y_j$$

But now our objective function value at optimality will not represent the number of rolls used, so we introduce an auxiliary variable, say t, and add

$$t = \sum_j y_j$$

7.1.1.3 Constraints

We have two types of constraints. The first is to ensure that customer demands are satisfied. So, across all rolls used, we must verify that we cut enough rolls of the required order quantity, say b_i,

$$\sum_{j \in K} x_{i,j} \geq b_i \quad \forall i \in N$$

The second type of constraints is to make sure that the consumer rolls we cut off the product roll do not exceed the width of that large roll, or, assuming that order i has width w_i, a percentage,

$$\sum_{i \in N} w_i x_{i,j} \leq 100 \quad \forall j \in K$$

But that is incomplete since we need to connect the x and y variables. No $x_{i,j}$ should be positive if the corresponding y_j is zero. We could introduce a large number of constraints or, realizing that we have encountered this situation before, simply modify the last constraint to read

$$\sum_{i \in N} w_i x_{i,j} \leq 100 y_j \quad \forall j \in K$$

7.1.1.4 Executable Model

Let's translate this into an executable model, as shown in Listing 7-1. It accepts a matrix D, exactly as in Table 7-1.

Listing 7-1. Cutting Stock Model with Pattern Search
(cutting_stock.py)

```
1  def solve_model(D):
2      s,n = newSolver('CuttinguStock', True), len(D)
3      k,b = bounds(D)
4      y = [s.IntVar(0,1,") for i in range(k[1])]
5      x = [[s.IntVar(0,b[i],") for j in range(k[1])] \
6          for i in range(n)]
7      w = [s.NumVar(0,100,") for j in range(k[1])]
8      nb = s.IntVar(k[0],k[1],")
9      for i in range(n):
10         s.Add(sum(x[i][j] for j in range(k[1])) >= D[i][0])
11     for j in range(k[1]):
12         s.Add(sum(D[i][1]*x[i][j] for i in range(n)) <= 100*y[j])
13         s.Add(100*y[j]-sum(D[i][1]*x[i][j] for i in range(n))==w[j])
14         if j < k[1]-1:
15             s.Add(sum(x[i][j] for i in range(n)) >= \
16                     sum(x[i][j+1] for i in range(n)))
17     Cost = s.Sum((j+1)*y[j] for j in range(k[1]))
18     s.Add(nb == s.Sum(y[j] for j in range(k[1])))
19     s.Minimize(Cost)
20     rc = s.Solve()
21     rnb = SolVal(nb)
22     return rc,rnb,rolls(rnb,SolVal(x),SolVal(w),D),SolVal(w)
```

At line 3 we call a routine bounds to find lower and upper bounds
on the numbers of rolls we will require and on the number of rolls of
each order that can fit in a roll. The upper bound on the number of rolls
used on line 4 to create as many y as we could possibly need (and also on
subsequent constraints related to each roll). The bound of the number of
rolls of each type is used at the next line to establish the range for each x:
on any given roll, we can have zero up to the number of elements that can
fit or the number required by the customer, hence the min expression.

Line 10 ensures that we satisfy every customer demand by summing over all rolls the elements of a given order. We could also have used an inequality here, namely ≥. The idea is that we may cut more than the customer demanded and then store these rolls in inventory, awaiting the next order. This sometimes makes sense, but is often better left out of the model. Once we have a solution that exactly satisfies customer demand, someone with an understanding of the situation and in possession of planning tools can decide how to best cut the leftover of the rolls. In either case, this will not change the total number of rolls used.

Line 12 ensures that the consumer rolls cut off a roll do not add up to more than 100% of the roll. And the next line, not a constraint, simply, computes the waste of each roll to help return a meaningful solution.

The loop starting at 14 breaks the symmetry of multiple solutions that are equivalent for our purposes: any permutation of the rolls. These permutations, and there are $K!$ of them, cause most solvers to spend an exorbitant time solving. With this constraint, we tell the solver to prefer those permutations with more cuts in roll j than in roll $j + 1$. The reader is encouraged to solve a medium-sized problem with and without this symmetry-breaking constraint. I have seen problems take 48 hours to solve without the constraint and 48 minutes with. Of course, for problems that are solved in seconds, the constraint will not help; it may even hinder. But who cares if a cutting stock instance solves in two or in three seconds? We care much more about the difference between two minutes and three hours, which is what this constraint is meant to address.

We could use, for the objective function, the simple sum of the rolls used variable, y, and pre-process the unused rolls. But we used a little trick to make each additional roll more "expensive" by including an ordinal factor. This guarantees that if, say the number of rolls is estimated to be between 11 and 14 and we end up using 12, they will be the first 12. There will be no "holes."

There are alternative objective functions. For example, we could have minimized the sum of the waste. This makes sense, especially if the demand constraint is formulated as an inequality. Then minimizing the sum of waste will spend more CPU cycles trying to find more efficient patterns that over-satisfy demand. This is especially good if the demand widths recur regularly and storing cut rolls in inventory to satisfy future demand is possible. Note that the running time will grow quickly with such an objective function.

Finally, we rearrange the solution to make sense for the caller. Instead of our decision variables x and y, we return a list of all rolls, with the cutting patterns and the waste of each.

Since we need bounds on the number of rolls and on the maximum number of cuts satisfying a given order on a single roll, Listing 7-2 implements a simple heuristic. The minimum is clearly the sum of all widths divided by the width of a roll, 100, as we assumed all widths to be entered as percentages. The upper bound is computed by a first fit heuristic: we add each roll, in order, to a roll until no more fits. Then we start a new roll. This is not brilliant, but it serves its purpose.

Listing 7-2. Cutting Stock Bounds Computation

```
1   def bounds(D):
2       n, b, T, k, TT = len(D), [], 0, [0,1], 0
3       for i in range(n):
4           q,w = D[i][0], D[i][1]
5           b.append(min(D[i][0],int(round(100/D[i][1]))))
6           if T+q*w <= 100:
7               T,TT = T+q*w,TT + q*w
8           else:
9               while q:
10                  if T+w <= 100:
11                      T,TT,q = T+w,TT+w, q-1
12                  else:
13                      k[1],T = k[1]+1,  0
14      k[0] = int(round(TT/100+0.5))
15      return k, b
```

Listing 7-3 reformats the solution to make it more meaningful to the caller. It returns an array containing for each roll used the percentage of waste incurred on that particular roll as well as the cut pattern used. Of course, the indicated cut pattern could be permuted with no change to the waste percentage. Output of our small example can be seen in Table 7-2.

Listing 7-3. Cutting Stock Model Solution Post-Process

```
1  def rolls(nb, x, w, D):
2    R,n = [],len(x)
3    for j in range(len(x[0])):
4      RR=[abs(w[j])]+[int(x[i][j])*[D[i][1]] for i in range(n) \
5                 if x[i][j]>0]
6    R.append(RR)
7    return R
```

Table 7-2. *Optimal Solution to the Cutting Stock*

Rolls	Waste 85.0	Pattern
0	5.0	{ 26; 23; 23; 23 }
1	16.0	{ 21; 21; 21; 21 }
2	4.0	{ 25; 25; 25; 21 }
3	4.0	{ 21; 21; 21; 33 }
4	1.0	{ 21; 26; 26; 26 }
5	1.0	{ 21; 26; 26; 26 }
6	0.0	{ 25; 21; 21; 33 }
7	0.0	{ 33; 33; 34 }
8	45.0	{ 25; 15; 15 }
9	1.0	{ 33; 33; 33 }
10	8.0	{ 25; 33; 34 }

7.1.2 Pre-Allocate Cutting Patterns

The previous approach is optimal but will not scale very well, even with our symmetry-breaking constraints. I will describe here a generally suboptimal approach that can be used to solve much larger instances.

The basic idea is to fix the cutting patterns and only optimize the number of rolls using those patterns while satisfying demand. Imagine, for instance, that we were given the distinct patterns of the last column of Table 7-2 in matrix A, as well as Table 7-1 in array D. Then we could have a decision variable y indexed by the patterns of A, representing how many rolls to cut according to that pattern. Symbolically, we would have the model

$$\min \sum_j y_j$$
$$A_{j,i} y_j \geq D_i \ \forall i$$
$$y_j \in [0,1,2,\ldots]$$

It seems simple enough, until one carefully considers the number of patterns. If we do not know ahead of time which patterns to use, then the simple option seems to be to list them all. How many are there? An astronomical number, even for small examples.

So, here is the brilliant idea: start with a certain set of patterns, optimize. Then, in a manner yet to be determined, find some "better" patterns to add. In all its generality, this is known to optimizers[1] as *column generation*. Repeat the optimization until we can no longer find "better" patterns or until we run out of time or are satisfied with the solution. Listing 7-4 implements the high-level approach.

---

[1]The expression column generation stemmed, in the minds of optimizers, as literal-minded as ever, from looking at the set of constraints as a matrix.

Listing 7-4. Cutting Stock Model Using Given Patterns
(cutting stock.py)

```
1   def solve_large_model(D):
2     n,iter = len(D),0
3     A = get_initial_patterns(D)
4     while iter < 20:
5       rc,y,l = solve_master(A,[D[i][0] for i in range(n)])
6       iter = iter + 1
7       a,v = get_new_pattern(l,[D[i][1] for i in range(n)])
8       for i in range(n):
9         A[i].append(a[i])
10      rc,y,l = solve_master(A,[D[i][0] for i in range(n)],True)
11      return rc,A,y,rolls_patterns(A, y, D)
12
13  def solve_master(C,b,integer=False):
14    t = 'Cuttingustockumasteruproblem'
15    m,n,u = len(C),len(C[0]),[]
16    s = newSolver(t,integer)
17    y = [s.IntVar(0,1000,") for j in range(n)] # right bound?
18    Cost = sum(y[j] for j in range(n))
19    s.Minimize(Cost)
20    for i in range(m):
21      u.append(s.Add(sum(C[i][j]*y[j] for j in range(n))
            >= b[i]))
22    rc = s.Solve()
23    y = [int(ceil(e.SolutionValue())) for e in y]
24    return rc, y, [0 if integer else u[i].DualValue() \
25                 for i in range(m)]
```

At line 4 we loop on the optimization a certain number of times. This is an easy termination criteria, which we can tune according to the size of the problem we are solving and how long we are willing to wait. There are better criteria, including looping until we have a true optimal solution, but they would lead us deep into the theory of optimization.[2]

The master problem is essentially the one described above: given the set of allowable patterns, do your best to minimize the number of rolls. There are two subtleties. The first one is that we solve the optimization problem as a linear program, not an integer program, even though we really want an integer solution, the number of rolls. We do this to gain speed. At the end, we simply round up the number of rolls, since obviously if 4.6 rolls will satisfy demand, then surely 5 rolls will also satisfy demand.

The second subtlety involves the information we need to find better patterns. Consider one imaginary example of a constraint of the form of line 21: for, say product roll 5, imagine we need 28 consumer rolls to satisfy demand. Given the patterns we have already, we might have that pattern 1 has this roll 3 times, pattern 3 has it 5 times, and pattern 10 has it once (and no other pattern has roll 5). So the constraint reads

$$3y_1 + 5y_3 + 1y_{10} \geq 28,$$

where the solution is y. What is the effect of changing the 28 by exactly 1 unit, keeping everything else constant? It will change the optimal solution by a small value, the marginal value of roll 5. We can do this, conceptually, for every roll and get each of their marginal values. By design, all solvers already have these marginal values computed; they are a side-effect of the solution techniques. So we simply request them at line 25 by the call for DualValues.

---

[2]The interested reader should look up "reduced cost" and "column generation of the cutting stock" to pursue these matters.

At the end, we reformat the solution to make it meaningful to the caller. We return an array containing for each roll used, the waste incurred on that particular roll, and the cut pattern used.

The model to find a new pattern for the master model to optimize over (Listing 7-5) uses the marginal value of each roll, provided by the solution to the master model above and maximizes the sum of the values times the number of occurrence of that roll in a pattern, while ensuring that the pattern stays within the total width of the large roll at line 7. This is a knapsack problem and will be solved very fast.

Listing 7-5. Cutting Stock Model (Getting a New Pattern)

```
1  def get_new_pattern(l,w):
2      s = newSolver('Cuttingustockuslave', True)
3      n = len(l)
4      a = [s.IntVar(0,100,") for i in range(n)]
5      Cost = sum(l[i]*a[i] for i in range(n))
6      s.Maximize(Cost)
7      s.Add(sum(w[i]*a[i] for i in range(n)) <= 100)
8      rc = s.Solve()
9      return SolVal(a), ObjVal(s)
```

We have left two elements undescribed: the initial patterns, and the reshaping of the solution to make it meaningful. From Listing 7-6, the initial patterns must be such that they will allow a feasible solution, one that satisfies all demands. We could be very creative here, or not. Considering the already complex model, let's go the latter route. Our initial patterns have exactly one roll per pattern, as obviously feasible as inefficient.

Listing 7-6. Cutting Stock Model (Getting a New Pattern)

```
1  def get_initial_patterns(D):
2    n = len(D)
3    return [[0 if j != i else 1 for j in range(n)]\
4            for i in range(n)]
5
6  def rolls_patterns(C, y, D):
7    R,m,n = [],len(C),len(y)
8    for j in range(n):
9      for _ in range(y[j]):
10       RR=[]
11       for i in range(m):
12         if C[i][j]>0:
13           RR.extend([D[i][1]]*int(C[i][j]))
14       w=sum(RR)
15       R.append([100-w,RR])
16    return R
```

A solution to our small example with this column generation approach is shown in Table 7-3. Note that it may use more rolls, but it actually cuts each roll very efficiently. It simply over-satisfies demands, unsurprising since we rounded up.

Table 7-3. *Possibly Sub-Optimal Solution to the Cutting Stock Using Column Generation*

Rolls	Waste 1	Pattern
0	0	{ 33; 33; 34 }
1	0	{ 25; 21; 21; 33 }
2	0	{ 25; 21; 21; 33 }
3	0	{ 25; 21; 21; 33 }
4	0	{ 25; 21; 21; 33 }
5	0	{ 25; 21; 21; 33 }
6	0	{ 26; 26; 33; 15 }
7	0	{ 26; 26; 33; 15 }
8	0	{ 25; 26; 26; 23 }
9	0	{ 25; 26; 26; 23 }
10	1	{ 21; 21; 23; 34 }

7.2 Non-Convex Trickery

When discussing piecewise objectives (in Section 3.1 of Chapter 3), I noted that if the function was not convex, then the approach suggested did not work. The situation was illustrated with a cost function modeling economies of scale; that is, unit prices decreased as we increased the number of units. For example, see Table 7-4.

Table 7-4. *Example of Non-Convex Piecewise Function*

(From	To]	Unit Cost	(Total Cost	Total Cost]
0	194	18	0	3492
194	376	16	3492	6404
376	524	14	6404	8476
524	678	13	8476	10478
678	820	11	10478	12040
820	924	6	12040	12664

If we tried to do something even as simple as minimizing this function, subject to $x \geq 250$ with our modal approach, we would get Table 7-5, which clearly does not solve our problem since

$$f(250) = 194 \cdot 18 + (250 - 194) \cdot 16$$
$$= 3492 + 56 \cdot 16$$
$$= 3492 + 896$$
$$= 4388.$$

Of course, if we had tried to maximize, it would have worked. The easy cases are minimizing convex functions and maximizing concave functions.[3] I will discuss here the hard cases.

Recall that the approach was to introduce decision variables λ_i, one per break in the function, to indicate which segment contains the optimal

---

[3]Optimizers working in engineering or applied mathematics traditionally minimize convex functions. There is a whole area of research appropriately called *convex analysis* devoted to the theory of such problems. In contrast, optimizers in business usually maximize concave functions. The theory is the same, but everything is upside down. Maybe we should call one group the optimizers and the other the pessimizers?

point and where on that segment the point is located (by the convex combination $x = \lambda_i P_i + \lambda_{i+1} P_{i+1}$). The model then became

$$\min \sum_{i=1}^{n} \lambda_i \sum_{j=1}^{i} \left(B_j - B_{j-1} \right) \times C_{j-1}$$

$$\sum_{i} \lambda_i = 1$$

$$x = \sum_{i} \lambda_i B_i$$

$$\lambda_i \in [0,1]$$

and other constraints.

Table 7-5. *Incorrect Solution to Non-Convex Piecewise Objective with* $x \geq 250$

Interval	0	1	2	3	4	5	6	Solution
δ_i	0.7294	0.0	0.0	0.0	0.0	0.0	0.2706	$\sum \delta = 1.0$
x_i	0	194	376	524	678	820	924	x=250.0
$f(x_i)$	0	3492	6404	8476	10478	12040	12664	Cost=3426

The problem with this model is that, even though at optimality the sum of the λ_i is one, we have two non-adjacent λ_i being non-zero. We must have two adjacent λ_i non-zero to determine which segment of the function to consider. We achieve this condition by introducing another set of binary variables, say $\delta_i \in \{0,1\}$, summing to one, and add the conditions

$$\lambda_0 \leq \delta_0$$
$$\lambda_1 \leq \delta_0 + \delta_1$$
$$\lambda_2 \leq \delta_1 + \delta_2$$
$$\lambda_3 \leq \delta_2 + \delta_3$$
$$\cdots$$
$$\lambda_{n-1} \leq \delta_{n-2} + \delta_{n-1}$$
$$\lambda_n \leq \delta_{n-1}$$

See what happens when one of the δ_i is one? Exactly two of the above inequalities, adjacent to each other, will have a right-hand side of one. Hence exactly two λ_i, adjacent, will be allowed to be non-zero.

This approach (two layers of binary variables) works with all integer solvers, but the situation of "At most two adjacent variables non-zero" occurs so often in practice that some solvers have special code to handle it. These variables are known as SOS2 (Special Ordered Set of type 2).[4] Which suggest the question: Is there a type 1? Indeed, there is: "Exactly one variable non-zero." But let's consider some useful generalizations with special cases as SOS1 and SOS2.

7.2.1 Selecting *k* Variables Out of *n* to Be Non-Zero

Consider a situation where we have a set of n variables $x_i \in [0, u_i]$ and we want to allow exactly k to be non-zero. For instance, if you are considering investing in various projects and are setting up a model to choose the best,

---

[4]Yet another example of the sadly unimaginative naming tradition of optimizers. Maybe this explains it. My Ph.D. advisor, said, only half in jest: "Do not name your algorithms with interesting names if you ever want them to be known by your name." The implication was that if one names his algorithms Alg-1 and Alg-2 or something equally pedestrian, colleagues will have no choice but to refer to them as Smith-star or Jonesrevised. Alas, such hope at posterity is belied by the use of SOS2 and other mutts of the same ilk.

say k of them. (If $k = 1$ we have the so-called SOS1 case.) We introduce n binary variables, λ_i, and add the constraints

$$x_i \le u_i \lambda_i \;\; \forall i, \tag{7.1}$$

$$\sum_i \lambda_i = k \tag{7.2}$$

We replace the equality in equation (7.2) with $\le$ if we want "at most" and with $\ge$ if we want "at least." Since this occurs regularly in modeling, let's create a general function that we might use within any given model. See Listing 7-7.

Listing 7-7. How to Select k Out of n Variables (`my or tools.py`)

```
1   def   k_out_of_n(solver,k,x,rel='=='):
2      n  =  len(x)
3      binary = sum(x[i].Lb()==0 for i in range(n)) == n and \
4             sum(x[i].Ub()==1 for i in range(n)) == n
5      if binary:
6        l = x
7      else:
8        l = [solver.IntVar(0,1,") for i in range(n)]
9        for i in range(n):
10         if x[i].Ub() > 0:
11           solver.Add(x[i]    <=    x[i].Ub()*l[i])
12         else:
13           solver.Add(x[i]    >=    x[i].Lb()*l[i])
14     S = sum(l[i] for i in range(n))
15     if rel == '==' or rel == '=':
16       solver.Add(S == k)
17     elif rel == '>=':
18       solver.Add(S >= k)
19     else:
20       solver.Add(S <= k)
21     return l
```

We craft Listing 7-7 to handle multiple cases. First, we need to single out binary variables because they do not need the additional layer of variables. We detect the binary case on line 4 by checking if all lower bounds are zero and all upper bounds are one. Any one variable not satisfying these conditions will set `binary` to `False`.

In the binary case, we simply rename parameter x to be l; in the other cases, we create the binary variable array l and then set the forcing bound of equation (7.1) at line 11 if $x \in [0,u]$ and correspondingly at line 13 if $x \in [l,0]$.

Finally, we set one of three relations on l, depending on whether the caller wants "exactly," "at most," or "at least" k variables selected. Note that the relation >= means "at least k variables are *allowed* to be non-zero." It does not mean that k variables *will* be non-zero.[5]

The reader might remember that in the first chapter (Section 2.1), when discussing variations, we left unsatisfied requirements of the form "If food 3 is used, then food 4 must not be (and vice versa)." This exclusive-or is now easily accommodated. Recall from Listing 2-1 that our food selection was using decision variable f. Then we can add one line to the model,

```
k_out_of_n(s, 1, [f[3],f[4]])
```

where foods 3 and 4 are inserted into an array to be passed to our newly-minted routine.

7.2.2 Selecting *k* Adjacent Variables Out of *n* to Be Non-Zero

If we want to generalize the non-zero adjacent constraint we used for the non-convex objective, we will need multiple layers of binary variables. Let's illustrate this by considering a set $x_i \in [0,u_i]$ of variables, out of which

---

[5]Though it is possible to model such constraints, it rarely makes much sense for continuous variables.

we want three adjacent to be non-zero. We introduce binary variables λ_i, δ_i and γ_i, satisfying the following:

$$x_0 \leq \lambda_0 u_0 \qquad\qquad \lambda_0 \leq \delta_0 \qquad\qquad \delta_0 \leq \gamma_0$$

$$x_1 \leq \lambda_1 u_1 \qquad\qquad \lambda_1 \leq \delta_0 + \delta_1 \qquad\qquad \delta_1 \leq \gamma_0 + \gamma_1$$

$$x_2 \leq \lambda_2 u_2 \qquad\qquad \lambda_2 \leq \delta_1 + \delta_2 \qquad\qquad \delta_2 \leq \gamma_1 + \gamma_2$$

$$x_3 \leq \lambda_3 u_3 \qquad\qquad \lambda_3 \leq \delta_2 + \delta_3 \qquad\qquad \delta_3 \leq \gamma_2 + \gamma_3$$

$$\ldots$$

$$x_{n-1} \leq \lambda_{n-1} u_{n-1} \qquad \lambda_{n-1} \leq \delta_{n-2} + \delta_{n-1} \qquad \delta_{n-1} \leq \gamma_{n-2}$$

$$x_n \leq \lambda_n u_n \qquad\qquad \lambda_n \leq \delta_{n-1}$$

$$\sum \lambda_i = 3 \qquad\qquad \sum \delta_i = 2 \qquad\qquad \sum \gamma_i = 1$$

To see how this set of constraints works, read backward from γ to λ. Only one of the γ_i is non-zero. This allows two adjacent δ_i to be non-zero, which, in turn, allows three adjacent λ_i to be non-zero. These last binary variables then correspond to the three adjacent xi that will be allowed to be non-zero. A nice recursive structure is implemented in Listing 7-8. We allow the caller some flexibility by accepting the number of variables selected to be zero or all of them; not that it makes sense in general, but it may help to structure a loop to include all boundary cases.

Listing 7-8. How to Select k Adjacent Variables Out of n to
Be Non-Zero

```
1  def sosn(solver,k,x,rel='<='):
2    def   sosnrecur(solver,k,l):
3      n = len(l)
4      d  =  [solver.IntVar(0,1,") for   _   in  range(n-1)]
5      for i in range(n):
6        solver.Add(l[i] <= sum(d[j] \
7          for j in range(max(0,i-1),min(n-2,i+1))))
8        solver.Add(k == sum(d[i] for i in range(n-1)))
9      return  d  if k  <=  1  else  [d,sosnrecur(solver,
       k-1,d)]
10   n  =  len(x)
11   if 0 < k < n:
12     l = k_out_of_n(solver,k,x,rel)
13     return l if k <= 1 else [l,sosnrecur(solver,k-1,l)]
```

The first layer of constraints is different from the others since the
variables might be continuous. This is handled at line 12 by calling the
function creating a layer of binary variables, setting bounds on each
continuous one, and returning the binary array. Then the recursive call to
sosnrecur, a private function, at line 13 implements the successive layers,
each one smaller by one variable. All the internal layers are returned to the
caller. The result of a very simple test choosing non-adjacent and adjacent
integers from a randomly created array to maximize their sum is shown in
Table 7-6.

Table 7-6. *Choosing the Largest Sum of k and of k Adjacent Variables*

Max Sum of	6	10	13	12	13	9	13	10	5
1/9			X						
Adjacent 1/9							X		
2/9					X		X		
Adjacent 2/9			X	X					
3/9			X		X		X		
Adjacent 3/9			X	X	X				
4/9			X	X	X		X		
Adjacent 4/9		X	X	X	X				
5/9			X	X	X		X	X	
Adjacent 5/9			X	X	X	X	X		
6/9		X	X	X	X		X	X	
Adjacent 6/9			X	X	X	X	X	X	
7/9		X	X	X	X	X	X	X	
Adjacent 7/9		X	X	X	X	X	X	X	
8/9	X	X	X	X	X	X	X	X	
Adjacent 8/9	X	X	X	X	X	X	X	X	
9/9	X	X	X	X	X	X	X	X	X
Adjacent 9/9	X	X	X	X	X	X	X	X	X

Let's now return to the non-convex objective function and see how to easily solve our problem. Executing Listing 7-9 on the same example (Table 7-4) will now produce the correct solution as can be seen in Table 7-7. Note that $\delta_i = 1$, allowing only λ_1 and λ_2 to be non-zero. So we are now in the correct segment of the piecewise function, between point 1 and point 2, and we can correctly determine both the solution x and the optimal value $f(x)$.

Listing 7-9. Piecewise Model for Non-Convex Function
(piecewise_ncvx.py)

```
1   def minimize_piecewise_linear(Points,B,convex=True):
2       s,n = newSolver('Piecewise', True),len(Points)
3       x = s.NumVar(Points[0][0],Points[n-1][0],'x')
4       l = [s.NumVar(0,1,'l[%i]' % (i,)) for i in range(n)]
5       s.Add(1 == sum(l[i] for i in range(n)))
6       d = sosn(s, 2, l)
7       s.Add(x == sum(l[i]*Points[i][0] for i in range(n)))
8       s.Add(x >= B)
9       Cost = s.Sum(l[i]*Points[i][1] for i in range(n))
10      s.Minimize(Cost)
11      rc  = s.Solve()
12      return  rc,SolVal(l),SolVal(d[1])
```

Table 7-7. *Correct Solution to Non-Convex Piecewise Objective with*
x ≥ 250

0	1	2	3	4	5	6	Solution
0.0	0.6923	0.3077	0.0	0.0	0.0	0.0	$\sum \lambda = 1.0$
0	194	376	524	678	820	924	x=250.0
0	1	0	0	0	0		$\sum \delta = 1$
0	3492	6404	8476	10478	12040	12664	f(x)=4388.00

7.2.3 Selecting *k* Constraints Out of *n*

A related trick is to select a certain number of constraints to be satisfied (and allowing others to be violated). Let's consider the case of one constraint, say

$$\sum_i a_i x_i \leq b, \tag{7.3}$$

where x is the decision variable. We may want to either raise an indicator variable if the constraint is satisfied or force the constraint if the indicator is raised:

$$\delta = 1 \Rightarrow \sum_i a_i x_i \leq b \tag{7.4}$$

or

$$\sum_i a_i x_i \leq b \Rightarrow \delta = 1 \tag{7.5}$$

This technique of associating a binary variable to the state of a constraint is known as *reifying* a constraint.[6]

Let's consider first the simpler equation (7.4). We need bounds

$$u_b := \max_x \sum_i a_i x_i - b,$$

$$l_b := \min_x \sum_i a_i x_i - b.$$

The bounds need not be exactly computed, although, as you will see, this is easily done. Any valid bound will work, with the usual caveat that in

---

[6]From *res* (genitive rei), Latin for object. The unusually creative nomenclature is due not to optimizers but to computer scientists working in the related (some would say adversarial) field of constraint programming.

order to avoid numerical difficulties, one should not use introduce "large" numbers. Armed with these parameters, we can add the constraint.

$$\sum a_i x_i \leq b + u_b (1 - \delta)$$

If $\delta = 0$, then the constraint is vacuous over the domain of x. If, on the other hand, $\delta = 1$, then the constraint must hold.

The other direction, equation (7.5), is neither as useful, nor as simple, but is an amusing exercise in translating logical expressions into algebraic ones.

First, let's formulate use the contrapositive of (7.5), namely

$$\delta = 0 \Rightarrow \sum_i a_i x_i \nleq b \text{ or}$$

$$\delta = 0 \Rightarrow \sum_i a_i x_i > b$$

In the case where a, b, and x are all integer variables, then the meaning of $\sum_i a_i x_i > b$ is clear. It means $\sum_i a_i x_i \geq b + 1$. The main difficulty occurs when x is a continuous variable. Then we need to decide the meaning of $>$ and it will be dependent on the problem we are modeling.

We need to decree that some e violation of the inequality is enough. If x represents wavelengths of visible light in meters, the value of e might be in the order of 10^{-9}. If x represent the money the US government spends on its military, then 105 might be fine. In any case, what we now want to implement is

$$\delta = 0 \Rightarrow \sum_i a_i x_i \geq b + \varepsilon \tag{7.6}$$

Once the modeler has decreed ε, we add

$$\sum a_i x_i \geq b + l_b \delta + \varepsilon (1 - \delta) \tag{7.7}$$

If $\delta = 0$, then this reduces to (7.6). If $\delta = 1$, then the lower bound comes into play and the constraint becomes vacuous. The case of "smaller than" is handled similarly or, even more easily, by multiplying everything by -1 and using the above. The case of equality is handled by transforming it into two inequalities.

Armed with these equations, we can now select k constraint out of n, by creating one indicator variable δ_i per constraint and using our previously defined k_out_of_n function on the δ. First, since we need bounds and we can easily set up a linear program to find them, let's do so. Listing 7-10 will find the tightest upper bound and lower bound on $\sum a_i x_i - b$ given a, x, and b.

Listing 7-10. How to Bound a Linear Constraint on a Box

```
1   from ortools.linear_solver import pywraplp
2   def bounds_on_box(a,x,b):
3       Bounds,n = [None,None],len(a)
4       s = pywraplp.Solver('Box',pywraplp.Solver.
        GLOP_LINEAR_PROGRAMMING)
5       xx = [s.NumVar(x[i].Lb(), x[i].Ub(),") for i in range(n)]
6       S = s.Sum([-b]+[a[i]*xx[i] for i in range(n)])
7       s.Maximize(S)
8       rc  = s.Solve()
9       Bounds[1] = None if rc != 0 else ObjVal(s)
10      s.Minimize(S)
11      s.Solve()
12      Bounds[0] = None if rc != 0 else ObjVal(s)
13      return Bounds
```

The reader might wonder why, at line 5, we create a copy of the provided parameter x instead of using the parameter itself. The reason is that x is attached to the caller's solver object, while the function bounds_on_box is creating a new solver instance. Worse yet, bounds_on_box may be called multiple times with the same x, each solver likely trying to bind x to different values. If we used the passed value, it would soon lead to an inconsistent model for the caller, or worse, produce nonsense without any indication of the origin of the problem. Hence the need for different variables.

After this detour to compute bounds on linear functions, we are in a position to implement the function to reify a constraint to a zero-one variable δ and enforce the constraint whenever δ is set. This is implemented in Listing 7-11.

The function `reify_force` accepts the required parameters defining the affine function $\sum a_j x_j - b$ in a, x, and b (note the sign). It also accepts three optional parameters. First, if the caller has another use for the indicator array, it may be created and passed in. If not, then it is created at line 5. In either case, it is returned. Second, the type of relation can be any of the three $\le, \ge$ or =. Finally, if the caller has bounds on the linear function, they may be passed in. It will avoid a call to our `bounds_on_box`. This would be necessary if, for example, the decision variable is not bounded by a box.

Listing 7-11. How to Reify a Constraint and How to Enforce It

```
1   def reify_force(s,a,x,b,delta=None,rel='<=',bnds=None):
2       # delta == 1 ---> a*x <= b
3       n = len(a)
4       if delta is None:
5           delta   =  s.IntVar(0,1,")
6       if bnds is None:
7           bnds = bounds_on_box(a,x,b)
8       if rel in ['<=','==']:
9           s.Add(sum(a[i]*x[i] for i in range(n))<=b+bnds[1]*
            (1-delta))
10      if rel in ['>=','==']:
```

```
11        s.Add(sum(a[i]*x[i] for i in range(n))>=b+bnds[0]*
          (1-delta))
12      return delta
```

We use the bounds_on_box function to find tight lower and upper bounds (assuming that the domain of x is tight) if the user does not provide such bounds. The reader should not be alarmed at the possibly large numbers of solvers that will be instantiated. Each instance is very small and runs extremely fast.

Finally, we add the appropriately modified constraint, either an implementation of equation (7.7) for a "less than or equal to" relation or the corresponding constraint for a "greater than or equal than." See Listing 7-12. In case the caller requires equality, we add both constraints since

$$\sum_j a_j x_j \le b \wedge \sum_j a_j x_j \ge b \Rightarrow \sum_j a_j x_j = b$$

Listing 7-12. How to Reify a Constraint and Raise an Indicator If It Is Satisfied

```
1  def reify_raise(s,a,x,b,delta=None,rel='<=',bnds=None,
   eps=1):
2    # a*x <= b ---> delta == 1
3    n = len(a)
4    if delta is None:
5      delta = s.IntVar(0,1,")
6    if bnds is None:
7      bnds = bounds_on_box(a,x,b)
8    if rel == '<=':
9      s.Add(sum(a[i]*x[i] for i in range(n)) \
10           >= b+bnds[0]*delta+eps*(1-delta))
11   if rel == '>=':
```

```
12       s.Add(sum(a[i]*x[i] for i in range(n)) \
13              <= b+bnds[1]*delta-eps*(1-delta))
14    elif rel == '==':
15       gm = [s.IntVar(0,1,") for _ in range(2)]
16       s.Add(sum(a[i]*x[i] for i in range(n)) \
17              >= b+bnds[0]*gm[0]+eps*(1-gm[0]))
18       s.Add(sum(a[i]*x[i] for i in range(n)) \
19              <= b+bnds[1]*gm[1]-eps*(1-gm[1]))
20       s.Add(gm[0] + gm[1] - 1 == delta)
21    return delta
```

The function `reify_raise` shares much structure, including the set of required and optional parameters with `reify _force`. The first difference is that, as discussed above, the caller must supply, in the case of continuous variables, the meaning of a violation, eps. The default is one, which works perfectly in the case of discrete variables.

The other difference is that we cannot rely of the parameter `delta` in all cases. We can in cases where the relation is $\leq$ or $\geq$, but not in the case of equality. The problem is that there are two ways in which an equality can fail: the left-hand side can be either greater or smaller than the right-hand side. This is why we introduce two other binary variables, `gm[0]` (really γ_0) and `gm[1]` (really γ_1), to reflect each type of violation:

$$\sum_j a_j x_j > b + \varepsilon \Rightarrow \gamma_0 = 1,$$
$$\sum_j a_j x_j < b - \varepsilon \Rightarrow \gamma_1 = 1$$

The γ array is then used to set δ using the following little trick. Since the γ cannot both be zero, they sum to either one or two, exactly when δ needs to be zero or one, hence $\gamma_0 + \gamma_1 - 1 = \delta$

As a final touch, we create the `reify` function that implements the *if and only if* condition that was implemented separately by `force` and `raise`. See Listing 7-13.

Listing 7-13. An Indicator Variable Set If and Only If a Constraint Is Satisfied

```
1  def reify(s,a,x,b,d=None,rel='<=',bs=None,eps=1):
2      # d == 1 <---> a*x <= b
3      return reify_raise(s,a,x,b,reify_force(s,a,x,b,d,rel,bs),
       rel, bs,eps)
```

7.2.4 Maximax and Minimin

To see an application of this trickery, recall that in Section 2.3.2.1 we left unsolved the problem of modeling the *maximax*,

$$\max_x \max_i \sum_j a_{i,j} x_j + b_i$$

subject to some constraints or the equivalently pernicious *minimin*. The technique is to first transform each affine function into a constraint of the form

$$\sum_j a_{i,j} x_j + b_i = z$$

We then transform each equality into a pair of inequalities, which we reify and to which we apply the disjunctive trick to enforce one of them. And finally we set the objective function to max z. (I did mention it was somewhat difficult to handle, but, with the routines we have developed in this section, it a matter of a few lines in Listing 7-14. As an example, we will solve the following,

$$\max_{x\in[2,5]} \max\{2x - 3, -2x + 12\},$$

which, graphically, is represented by Figure 7-2 where the two functions are displayed and the maximum is in a thicker line. Note that this is decidedly a non-convex objective. Note also that one could solve such a simple problem differently, but not that much more efficiently. For example, one might evaluate all functions on the vertices of the polyhedron. But in order to do that, one needs to find the vertices. To find those, one needs to either solve an exponential number of linear programs or an exponential number of linear system of equations. In the worst case, our approach can theoretically require as much work, but in practice, it never does.[7]

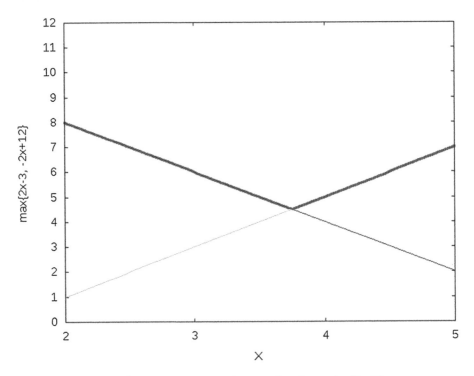

Figure 7-2. *max{2x − 3, −2x + 12} over the domain [2, 5]*

---

[7]The interested reader needs to research the branch-and-bound technique of integer programming.

Listing 7-14. How to Solve Maximax Problems (`my or tools.py`)

```
1  def maximax(s,a,x,b):
2      n = len(a)
3      d = [bounds_on_box(a[i],x,b[i]) for i in range(n)]
4      zbound = [min(d[i][0] for i in range(n)), max(d[i][1] \
5              for i in range(n))]
6      z = s.NumVar(zbound[0],zbound[1],")
7      delta = [reify(s,a[i]+[-1],x+[z],b[i],None,'==') \
8              for i in range(n)]
9      k_out_of_n(s,1,delta)
10     s.Maximize(z)
11     return z,delta
```

The function `maximax` receives the solver `s` to which to add the maximax constraints, along with matrix a and arrays x and b, representing the n affine functions $\sum_j a_{i,j} x_j + b_i$ for $i \in [0, n-1]$. We create the additional variable z which will become the objective to maximize. To set meaningful bounds on z, we use `bounds_on_box` to find the minima and maxima of all the affine functions on the domain of x. We use the minimum and maximum of those for our bounds.

Each of these functions is then set equal to z and reified on a corresponding `delta[i]` so that $\delta_i = 1 \Leftrightarrow \sum_j a_{i,j} x_j + b_i = z$. Finally, we force exactly one of the `delta[i]` to be one, or equivalently, one of the affine function constraints to be active. We set the objective and return both the objective and the array of indicators. This will provide the caller with all the necessary information at optimality.

The solution to our small example is clearly $x = 2$ with objective value $-2x + 12 = 8$. Indeed, running Listing 7-14 returns 8.0, [0, 1] telling us that the second affine function is the active one.

7.3 Staff Scheduling

Staff scheduling is not one problem but a vast array of problems, each with its own set of requirements and quirks. I will discuss one interesting variation. It involves assigning instructors to course sections. The main characteristic of this problem, which makes it interesting, is the handling of instructor preferences.

Here is the generic situation: course sections have been assigned meeting times during the week. For instance, MOR142[8] is worth three credits; Section 1 meets Monday, Wednesday, and Friday from 9:00 to 10:00 while section 2 meets Tuesday and Thursday from 9:00 to 10:30. Each of these sections needs one instructor. There are dozens of these sections for various courses, each with assigned times and credits, requiring instructors.

On the other hand, we have a set of instructors, some full time who will teach a fixed number of credits, and some part time who can teach up to a certain number of credits. Moreover, no instructor has managed to clone himself in order to teach two parallel sections offered at the same times. These are hard constraints.

In addition, each instructor has expressed preferences (or dislikes) for certain courses (course preferences), and days or times (we will call these preference sets). For instance, we could have a set of "Sections offered on Monday, Wednesday, Friday" and another one of "Sections meeting at night." The instructor could give a weighted thumbs up (or down) to each set.

If these were all the required constraints, then the model would simply be that of an assignment problem. But real scheduling is never, ever, as trivial as assignments. So let's consider one additional constraint.

This is where it gets interesting: each instructor has also indirectly expressed preferences (or dislikes) to pairs of sections. For instance there could be an abstract pair "A section meeting on Monday night and another one meeting on Tuesday morning" or "A section meeting, followed within an hour by another meeting." One can easily imagine why an instructor might want to avoid (or might prefer) to have back-to-back section meetings.

---

[8]Mathematics of Operations Research, the umbrella title for the topics of this book.

To illustrate, let's assume an instance where all these preferences and preferences pairs have been processed and expressed in their simplest form in Table 7-8 for the sections, Table 7-9, and Table 7-11. A good deal of preprocessing might be required to extract the data and format it in these tables, but this is not currently our concern.

In Table 7-8, the first column is an ordinal indicating the section while the second indicates the course. In our example, the first two rows might correspond to the first two sections of MOR142. The third column is an indication of the time (time 12 might be Monday, Wednesday, and Friday at 9:00).

Table 7-8. *List of Sections Offered*

Id	Course Id	Meeting Time
0	0	12
1	0	19
2	1	11
3	1	12
4	2	11
5	3	16
6	3	2
7	3	7
8	4	17
9	5	1
10	5	20
11	5	20
12	6	13
13	6	4
14	6	1

(continued)

Table 7-8. (*continued*)

Id	Course Id	Meeting Time
6	3	2
7	3	7
8	4	17
9	5	1
10	5	20
11	5	20
12	6	13
13	6	4
14	6	1

In Table 7-9 the first column is the ordinal identification of the instructor, followed by the course load range. The third column has the preferences (positive integer) or dislikes (negative integer) of the instructor for each of the courses, in the order of Table 7-8. The fourth column holds the preferences corresponding to the sets to which sections belong in the order of Table 7-10. The last column is the preference for the pairs found in Table 7-11.

Table 7-9. *List of Preferences of Each Instructor*

Id	Load	Course Weights	Set Weights	Pair Weights
0	{ 2; 3 }	{ 0; 2; 0; 0; 0; 0; -4 }	{ 0; 0; 7; -5; -6; 0 }	{ 0; 0 }
1	{ 2; 2 }	{ 0; 3; 2; 0; 0; 10; 0 }	{ 0; 0; 0; 8; 4; 9 }	{ 0; 8 }
2	{ 1; 3 }	{ 2; -2; 2; 0; 8; -2; 2 }	{ 0; 0; 0; 0; 0; 9 }	{ 0; 0 }
3	{ 1; 2 }	{ 3; 0; 0; 0; 9; -2; -4 }	{ 0; 7; 9; 0; 0; 0 }	{ 0; 0 }
4	{ 2; 2 }	{ 0; -10; 1; 0; 0; 0; -6 }	{ 0; -1; 3; 10; -6; 0 }	{ 0; -7 }

Table 7-10 lists the sections corresponding to each preference set.

Table 7-10. *List of Preferences Sets*

Id	Sections
0	{ 0; 7; 8; 9; 11; 14 }
1	{ 0; 1; 2; 7; 9; 11; 12 }
2	{ 0; 2; 5; 6; 10; 11 }
3	{ 1; 3; 6; 8; 13; 14 }
4	{ 1; 4; 5; 7; 10 }
5	{ 0; 2; 5; 7; 8; 11; 12 }

Finally, Table 7-11 lists the sections corresponding to each preference set.

Table 7-11. *List of Preferences Pairs*

Id	Section Pairs
0	{ (3 7); (9 12); (10 14) }
1	{ (10 11); (11 13); (11 14) }

7.3.1 Constructing a Model

We will describe the model in stages.

7.3.1.1 Decision Variables

What we need to decide in this problem is which instructor to assign to which section. So, clearly by now, the decision variable could be binary, indexed by the set of instructors I and the set of sections S as

$$x_{i,j} \in \{0,1\} \quad \forall i \in I; \forall j \in S,$$

where $x_{13,61} = 1$ indicates that instructor id 13 is assigned to section id 61.

We will likely need a considerable number of auxiliary variables to construct a readable model. Let's start on the constraints and introduce the auxiliaries as need be.

7.3.1.2 Constraints

Each section needs to be assigned at most one instructor,

$$\sum_i x_{i,j} \leq 1 \quad \forall j \in S$$

Each instructor must be assigned a number of courses within a certain range, say $[L_i, U_i]$,

$$L_i \leq \sum_i x_{i,j} \leq U_i \quad \forall j \in I$$

Now for the no-cloning constraint, that is, each instructor can be busy with at most one section per meeting time. Assume that the set of meeting times is T, then

$$\sum_{j:T_j=t} x_{i,j} \leq 1 \quad \forall j \in T,$$

where T_j is the meeting time of section j.

7.3.1.3 Objective

The objective needs to maximize the preference weights of all the instructors. We will split the objective into three terms, one for the weights of instructor i on course c ($wc_{i,c}$), on preference set s ($ws_{i,s}$) and on preference pair p ($wp_{i,p}$).

For the courses, this is simple. Assuming that the set of sections S is partitioned into subsets S_c for the sections of course id c, then the contribution of the course preference weights is

$$WC = \sum_{c \in C} \sum_{i \in I} wc_{i,c} \sum_{j \in S_c} x_{i,j} \tag{7.8}$$

The contribution of the set preference weights is also fairly straightforward. We need to sum over all the preference sets and all the instructors the product of the weight an instructor puts on a set with the sum over all sections of the indicator of set membership of that section and the indicator of assignment of that section to the instructor,

$$WR = \sum_{s \in S} \sum_{i \in I} ws_{i,s} \sum_{j \in R_s} x_{i,j},$$

where R_s is the last column of Table 7-10.

Now for the more interesting weights, on pairs. Let's look at a specific example. Say that pair id 4 indicates consecutive meetings and that it includes a pair of sections 2 and 5. Also, instructor 13 has put a weight of -15 on such consecutive pairs. Then if we assign sections 2 and 5 to instructor 13, we need to add -15 to the objective value. So we need an indicator for "Section 2 and 5 are assigned to instructor 13." Let's call this indicator $z_{13,4}$. According to our model, we know that $x_{13,2}$ and $x_{13,5}$ will be one. How can we set $z_{13,4}$ if and only if both are one? By

$$x_{13,2} + x_{13,5} - 1 \leq z_{13,4},$$
$$z_{13,4} \leq x_{13,2},$$
$$z_{13,4} \leq x_{13,5}.$$

The first inequality raises z when both x are one. The last two lower z to zero when either x is zero.

Now, in all generality, we obtain, assuming sets of pairs P_p as in the last column of Table 7-11,

$$x_{i,s_1} + x_{i,s_2} - 1 \leq z_{i,p}, \; i \in I, (s_1, s_2) \in P_p, \tag{7.10}$$

$$x_{i,s_1} \geq z_{i,p}, \tag{7.11}$$

$$x_{i,s_2} \geq z_{i,p}. \tag{7.12}$$

An alternative approach, using what we have previously developed in the section on non-convex tricks (Section 7.2), is to use the reify high-level constraint which will implement

$$x_{i,s_1} + x_{i,s_2} \geq 2 \Leftrightarrow z_{i,p} = 1$$

We can now sum over all preference pairs and all instructors the product of the weight and the indicator,

$$WP = \sum_p \sum_i z_{i,p} wp_{i,p} \tag{7.13}$$

We now have the complete objective function as

$$\max WC + WS + WP.$$

7.3.1.4 Executable Model

Let's translate this into an executable model. See Listing 7-15.

Listing 7-15. Staff Scheduling Model (staff_scheduling.py)

```
1   def solve_model(S,I,R,P):
2       s = newSolver('StaffuScheduling',True)
3       nbS,nbI,nbSets,nbPairs = len(S),len(I),len(R),len(P)
4       nbC,nbT = S[-1][1]+1,1+max(e[2] for e in  S)
5       x=[[s.IntVar(0,1,") for _ in range(nbS)] for _ in
        range(nbI)]
6       z=[[[s.IntVar(0,1,") for _ in range(len(P[p][1]))] \
7             for p in range(nbPairs)] for _ in range(nbI)]
8       for j in range(nbS):
9         k_out_of_n(s,1,[x[i][j] for i in range(nbI)],'<=')
10      for i in range(nbI):
11        s.Add(sum(x[i][j] for j in range(nbS)) >= I[i][1][0])
12        s.Add(sum(x[i][j] for j in range(nbS)) <= I[i][1][1])
13        for t in range(nbT):
```

```
14          k_out_of_n(s,1,
15              [x[i][j] for j in range(nbS) if S[j][2]==t],'<=')
16      WC=sum(x[i][j] * I[i][2][c] for i in range(nbI) \
17          for j in range(nbS) for c in range(nbC) if S[j][1] == c)
18      WR=sum(I[i][3][r] * sum(x[i][j] for j in R[r][1]) \
19          for r in range(nbSets) for i in range(nbI))
20      for i in range(nbI):
21        for p in range(nbPairs):
22          if I[i][4][p] != 0:
23            for k in  range(len(P[p][1])):
24                xip1k0,xip1k1=x[i][P[p][1][k][0]],x[i][P[p][1][k][1]]
25                reify(s,[1,1],[xip1k0,xip1k1],2,z[i][p][k],'>=')
26      WP = sum(z[i][p][k]*I[i][4][p] for i in range(nbI) \
27                for p in range(nbPairs) for k in range(len(P[p][1])) \
28                if I[i][4][p] != 0)
29      s.Maximize(WC+WR+WP)
30      rc,xs = s.Solve(),ss_ret(x,z,nbI,nbSets,nbS,nbPairs,I,S,R,P)
31      return  rc,SolVal(x),xs,ObjVal(s)
```

The function solve _model receives in S the section data in the form
of Table 7-8; in I the instructor data in the form of Table 7-9; in R the
preference sets data in the form of Table 7-10; and in P the preference pairs
data in the form of Table 7-11.

The decision variable x on line 5 is declared as a two-dimensional
array, indexed by section and instructor id. On the following line, the
auxiliary variable z indexed by instructor id, preference pair id, and the
ordinal of the pairs of sections within a preference pair will be one if we
assign one of the pairs to the instructor.

The loop on line 8 ensures that each section has at most one instructor.
We are assuming here that there are more sections to teach than
instructors to teach them. The subloop on the set of meeting times at 13
ensures that no instructor is required to be at two places at the same time.

The loop on 10 is an availability constraint; it ensures that each
instructor teaches as many courses as she is supposed to teach.

At that point we have all we need to compute two of the objective terms: the weighted course preference at line 17 and the weighted set preference at line 19. These correspond to equations (7.8) and (7.9).

To implement equation (7.13) at line 28, we need to loop over all instructors, all sets of pairs, and all pairs (the triple loop at line 20), and reify pairs of sections assigned to an instructor. We do this if and only if the instructor used a non-zero weight on such pairs. There is no point in adding such complex constraints if the net effect on the objective function is zero.

Listing 7-16. Staff Scheduling Meaningful Solution

```
1   def ss_ret(x,z,nbI,nbSets,nbS,nbPairs,I,S,R,P):
2       xs=[]
3       for i in range(nbI):
4         xs.append([i,[[j,(I[i][2][S[j][1]],\
5           sum(I[i][3][r] for r in range(nbSets) if j in R[r][1]),
6           sum(SolVal(z[i][p][k])*I[i][4][p]/2
7               for p in range(nbPairs) for k in range(len(P[p][1]))
8                 if j in P[p][1][k]))] for j in range(nbS) \
9                 if SolVal(x[i][j])>0]])
10      return xs
```

Finally, after we solve it, we massage the solution to return to the caller a meaningful answer constructed by Listing 7-16: a list, indexed by instructor, containing all of his assigned sections with, for verification purposes, the three weights that participated in this assignment as seen in Table 7-12. The weight on preference pairs is split in two for the two sections that triggered this weight to participate in the optimal value. This is useful information for the user to lift the veil on the optimization model performance.[9]

---

[9]And priceless during the inevitable discussions with staff complaining vociferously that their preferences were not met. A modeler's work is incomplete until the user accepts the solution.

Table 7-12. *Optimal Solution to the Staff Scheduling*

Instructor	Section:(WC WR WP)		
0	2 : (2 7 0)	5 : (0 1 0)	10 : (0 1 0)
1	11 : (10 9 4)	14 : (0 8 4)	
2	7 : (0 9 0)	8 : (8 9 0)	12 : (2 9 0)
3	0 : (3 16 0)	1 : (3 7 0)	
4	6 : (0 13 0)	13 : (-6 10 0)	

7.3.2 Variations

Without modifying the overall structure of the model above, a number of variations and additional constraints are possible.

- In a typical department, not all instructors are qualified to teach all courses. A "Qualified" Boolean could be attached to each Instructor-Course pair to prevent some assignments. This is simple to accommodate by setting to zero the corresponding decision variables for all the sections of the course.

- The department might have a policy whereby a subset of the instructors, say tenured professors, are forced to teach one lower-level course per semester, no matter what their preferences states. This would be implemented as a k-out-of-n type constraint.

- For certain courses with a large number of sections, the department might want at least one tenured professor teaching a section, while all other sections could be instructed by adjuncts. Again, this is a k-out-of-n type constraint, with the appropriate set.

7.4 Sports Timetabling

By sports timetabling, I mean the construction of a schedule of games for a league.[10] If you do not care about spectator sports, read on anyway because the problem is interesting, very difficult, and leads us into the fascinating and complex area of *relaxation tightening*, which can be applied to other complex problems.

Here is the generic problem we will try to solve: the league has a number of divisions, each with a certain number of teams. The league specifies the number of times during a complete season that each team must face each other team of the same division and of each other division as well as the maximum number of games in a week each team will play.

For a simple instance, see Table 7-13. The "Intra" parameter is the number of times each team faces each other team of the same division. The "Inter" is for each team of other divisions. "G/W" is the number of games per week of each team and "Weeks" is the number of weeks of the season.

Table 7-13. *Example of Sports Timetabling Data*

(Intra Inter G/W Weeks)	{ 2; 1; 1; 19 }
Division 0 teams	{ 0; 1; 2; 3; 4; 5; 6 }
Division 1 teams	{ 7; 8; 9; 10; 11; 12; 13 }

7.4.1 Constructing a Model

We will describe the model in stages.

7.4.1.1 Decision Variables

What is the end result of this model? A calendar of sorts, something that will display that on Week 5, for instance, Teams 1&3, Teams 2&7, etc. are facing each other (and this, for every week of the season). How can we

---

[10]Think NBA if you are US American, NHL if Canadian, ARL if Australian, or EPL if you are the colonizer of the previous three.

encode this information? One possibility is for a three-dimensional binary variable $x_{i,j,w}$ where i and j are team indices $(i < j)$ and w is a week index. The interpretation is, for example, if $x_{2,5,13}$ is one, then teams 2 and 5 meet during week 13.

Does this strike the reader as profoundly inefficient? **It is!**

It has the redeeming value that some of the constraints will be wonderfully simple to express. If this model works, we are done. If not, then we can try harder. Let's pursue this avenue further.

7.4.1.2 Constraints

The first constraint is that we have a fixed number (say n_A) of intra-division games between teams (say T_d) of each division, or

$$\sum_w x_{i,j,w} = n_A \quad \forall i \in T_d; \forall j \in T_d; i < j; \forall d \in D$$

The inter-division constraint is similar. For number of games n_R,

$$\sum_w x_{i,j,w} = n_R \quad \forall i \in T_d; \forall j \in T_e; \forall d \in D; \forall e \in D; d < e$$

The number of games per week that a team plays is actually an upper bound. Imagine a simple boundary case of one division with three teams and one game per week. One of the teams cannot possibly play. Therefore, we need an inequality. For number of games n_G,

$$\sum_{i<j} x_{i,j,w} + \sum_{j<i} x_{j,i,w} \leq n_G \quad \forall i \in T; \forall w \in W$$

Notice the two sums. Since we fix team i, we must look at the games with teams with both larger and smaller ordinals.

7.4.1.3 Objective Function

This problem is complex enough that even feasibility is difficult. Besides, the possible objective functions likely vary considerably with leagues. For the sake of illustration, let's assume that we want, as much as possible, to push intra-divisional matches towards the end of the calendar. The later, the better.

Let's consider two teams of the same division, i and j. If they face each other during week w, then the variable $x_{i,j,w}$ will be one. How can we put a weight on this according to "lateness?" We could multiply by w. This leads us to the following objective,

$$\sum_{w \in W} \sum_{d in D} \sum_{i \in T_d} \sum_{j \in T_d | i < j} w x_{i,j,w}$$

Unfortunately, this objective performs rather badly sometimes. For our purposes, all solutions that have intra-divisional games towards the end are good. There is no reason to favor the last week over the second-to-last week. So let's do some calculations and compute the number of weeks required for intra-divisional games. For n_A games and $|T_d|$ teams in a division and a maximum of n_G games, we get that we need n_w weeks,

$$n_w = \frac{|T_d| n_A}{n_G}$$

So if we assign one for intra-divisional games in the last n_w weeks, and zero otherwise, we obtain the objective.

$$\sum_{w=|W|-n_w}^{|W|} \sum_{d \in D} \sum_{i \in T_d} \sum_{j \in T_d | i < j} x_{i,j,w},$$

which performs much better, usually.[11] Notice that this computation of the required number of weeks is not always correct; it can be off by one but it serves our purposes.

---

[11]For the theoretically-minded, the primal-dual gap is smaller; optimality detection is easier.

7.4.1.4 Executable Model

Let's translate this into an executable model. It will receive a list of divisions, each containing the teams of that division. It would be simpler if all divisions had the same number of teams, but that is never the case.

It also accepts a list of parameters, the number of intra-divisional games, nbIntra; of inter-divisional games, nbInter; of games per week for a team, nbPerWeek (note that this has to be a maximum, not a strict constraint); and the number of weeks of the season, nbWeeks. See Listing 7-17.

Listing 7-17. Sports Timetabling Model (`sports timetabling.py`)

```
1   def solve_model(Teams,params):
2     (nbIntra,nbInter,nbPerWeek,nbWeeks) = params
3     nbTeams = sum([1 for sub in Teams for e in sub])
4     nbDiv,Cal = len(Teams),[]
5     s = newSolver('Sportsuschedule', True)
6     x = [[[s.IntVar(0,1,") if i<j else None
7           for _ in range(nbWeeks)]
8           for j in range(nbTeams)] for i in range(nbTeams-1)]
9   for Div in Teams:
10    for i in Div:
11       for j in Div:
12          if i<j:
13             s.Add(sum(x[i][j][w] for w  in range(nbWeeks))
              ==nbIntra)
14    for d in range(nbDiv-1):
15     for e in range(d+1,nbDiv):
16       for i in Teams[d]:
17          for j in Teams[e]:
18             s.Add(sum(x[i][j][w] for w  in range(nbWeeks))
              ==nbInter)
19    for w in range(nbWeeks):
```

```
20      for i in range(nbTeams):
21          s.Add(sum(x[i][j][w] for j in range(nbTeams) if i<j)
+
22              sum(x[j][i][w] for j in range(nbTeams) if j<i )\
23              <=nbPerWeek)
24      Value=[x[i][j][w] for Div in Teams for i in Div for j in
        Div \
25          for w in range(nbWeeks-len(Div)*nbIntra//nbPerWeek,
            nbWeeks) \
26          if i<j]
27      s.Maximize(sum(Value))
28      rc  = s.Solve()
29      if rc == 0:
30        Cal=[[(i,j) \
31              for i in range(nbTeams-1) for j in
                range(i+1,nbTeams)\
32              if SolVal(x[i][j][w])>0] for w in range(nbWeeks)]
33      return rc,ObjVal(s),Cal
```

Line 7 declares our decision variables. Note that this is a list of lists of lists. The first dimension is one less than the number of teams since we only will consider matches i vs j where $i < j$. The second dimension is the number of teams but note that half of the entries (below the diagonal) will never be used so we set them to None. The last dimension is the number of weeks.

Line 9 starts a loop to set the number of intra-division games. We loop on each division, and then on every pair (i, j) of teams in the division, respecting the $i < j$ condition to only use the upper triangle.

Similarly for the loop starting at 14 where we loop on each division, then on every division with a larger ordinal, then for every pair of teams, each in one of the two divisions.

Finally, after we solve it, we massage the solution to return to the caller a meaningful result: a list of matches, indexed by the week's ordinal. For our small instance, a result is shown in Table 7-14.

Table 7-14. *Optimal Solution to the Sports Timetabling*

Week	Matches						
0	0 vs 12	1 vs 11	2 vs 7	3 vs 13	4 vs 9	5 vs 8	6 vs 10
1	0 vs 9	1 vs 10	2 vs 13	3 vs 11	4 vs 8	5 vs 12	6 vs 7
2	0 vs 11	1 vs 12	2 vs 8	3 vs 7	4 vs 13	5 vs 10	6 vs 9
3	0 vs 13	1 vs 7	2 vs 10	3 vs 9	4 vs 12	5 vs 11	6 vs 8
4	0 vs 8	1 vs 13	2 vs 9	3 vs 12	4 vs 10	5 vs 7	6 vs 11
5	0 vs 2	1 vs 4	3 vs 5	6 vs 13	7 vs 12	8 vs 10	9 vs 11
6	0 vs 3	1 vs 4	2 vs 6	5 vs 9	7 vs 13	8 vs 11	10 vs 12
7	0 vs 4	1 vs 8	2 vs 3	5 vs 6	7 vs 13	9 vs 10	11 vs 12
8	0 vs 1	2 vs 12	3 vs 6	4 vs 5	7 vs 8	9 vs 11	10 vs 13
9	0 vs 5	1 vs 6	2 vs 4	3 vs 10	7 vs 11	8 vs 12	9 vs 13
10	0 vs 6	1 vs 3	2 vs 5	4 vs 11	7 vs 10	8 vs 13	9 vs 12
11	0 vs 1	2 vs 6	3 vs 4	5 vs 13	7 vs 11	8 vs 12	9 vs 10
12	0 vs 2	1 vs 5	3 vs 8	4 vs 6	7 vs 9	10 vs 11	12 vs 13
13	0 vs 6	1 vs 9	2 vs 3	4 vs 5	7 vs 12	8 vs 11	10 vs 13
14	0 vs 5	1 vs 6	2 vs 11	3 vs 4	7 vs 10	8 vs 13	9 vs 12
15	0 vs 4	1 vs 3	2 vs 5	6 vs 12	7 vs 9	8 vs 10	11 vs 13
16	0 vs 7	1 vs 2	3 vs 5	4 vs 6	8 vs 9	10 vs 11	12 vs 13
17	0 vs 3	1 vs 2	4 vs 7	5 vs 6	8 vs 9	10 vs 12	11 vs 13
18	0 vs 10	1 vs 5	2 vs 4	3 vs 6	7 vs 8	9 vs 13	11 vs 12

This approach works for smallish instances but will not scale very well as the reader can attest by trying larger instances. (Try something the size of a professional league and be prepared to wait a while for a solution.) The problem stems from the interaction between the model feasible solution space and the techniques used by integer programming solvers to find optimal solutions. Solvers typically work by fixing some of the variables in the model and then solving for the others by letting them take on a fractional solution: iterating multiple times. For our model, this relaxation is fairly weak. I will not go into the details of why, but you will see how to fix it, once you realize that the solvers are slow to solve. The key insight is that we can easily add more constraints.

Here is an example. Imagine an instance with one game per week and consider three teams, say i, j, and k. If the decision variable is allowed to take on fractional values at some point during the execution, then for a given week w it could happen that the optimal solution satisfies the following:

$$x_{i,j,w} = \frac{1}{2} \quad x_{i,k,w} = \frac{1}{2} \quad x_{j,k,w} = \frac{1}{2}$$

Note that this solution is allowed by the constraint that says "One game per week per team" since

$$x_{i,j,w} + x_{i,k,w} = 1,$$
$$x_{i,j,w} + x_{j,k,w} = 1,$$
$$x_{i,k,w} + x_{j,k,w} = 1$$

But we know that this is not a valid solution since the sum of the three variables must not exceed one. If i and j face each other, then k cannot face either of them; similarly for the pair (i, k) and for (j, k). Therefore, knowing that there is only one game per week, we could add a constraint for every triple of team i, j, k for every week w,

$$x_{i,j,w} + x_{i,k,w} + x_{j,k,w} \leq 1$$

Note that this constraint, if the variables are integers, is redundant. But it is a valid constraint nevertheless and it is useful for fractional values, which is happening internally in the solver. We can also consider tuples of four or even five teams, and number of games per week of two or three. See Table 7-15 for the bounds given small numbers of teams and of games per week. Note that many of these bounds will never be violated by fractional solutions so they are not very useful for our purposes.

Table 7-15. *Bounds on Small Sums of Tuples of Decision Variables*

Nb of Teams	Nb of Games Per Week	Bound on Sum
3	1	1
	2	3
4	1	2
	2	4
	3	6
5	1	2
	2	5
	3	7
	4	9

The number of those additional constraints grows fast. The approach will therefore add a considerable number of constraints to the model. If that causes solvers to slow down unacceptably, an alternative is the approach we used to solve the TSP: adding only the constraints that we need. We would do that by solving the relaxation, looking for tuples violating the bound, and adding them. This is the aim of Listing 7-18. It is an example of how easily one can add relaxation-tightening constraints to a model written with OR-Tools.

Listing 7-18. Sports Timetabling with Additional Cuts

```
1   def  solve_model_big(Teams,params):
2     (nbIntra,nbInter,nbPerWeek,nbWeeks) = params
3     nbTeams = sum([1 for sub in Teams for e in sub])
4     nbDiv,cuts  =  len(Teams),[]
5     for iter in range(2):
6       s = newSolver('Sportsuschedule', False)
7       x = [[[s.NumVar(0,1,") if i<j else None
8              for _ in range(nbWeeks)]
9              for j in range(nbTeams)] for i in range
               (nbTeams-1)]
10      basic_model(s,Teams,nbTeams,nbWeeks,nbPerWeek,nbIntra,\
11                  nbDiv,nbInter,cuts,x)
12      rc = s.Solve()
13      bounds = {(3,1):1, (4,1):2, (5,1):2, (5,3):7}
14      if nbPerWeek <= 3:
15        for w in range(nbWeeks):
16          for i in range(nbTeams-2):
17            for j in range(i+1,nbTeams-1):
18              for k in range(j+1,nbTeams):
19                b = bounds.get((3,nbPerWeek),1000)
20                if sum([SolVal(x[p[0]][p[1]][w]) \
21                      for p in pairs([i,j,k],[])])>b:
22                  cuts.append([[i,j,k],[w,b]])
23                  for l in range(k+1,nbTeams):
24                    b = bounds.get((4,nbPerWeek),1000)
25                    if sum([SolVal(x[p[0]][p[1]][w]) \
26                          for p in pairs([i,j,k,l],[])])>b:
27                      cuts.append([[i,j,k,l],[w,b]])
28                      for m  in range(l+1, nbTeams):
29                        b = bounds.get((5,nbPerWeek),1000)
```

```
30                    if sum([SolVal(x[p[0]][p[1]][w]) \
31                          for p in pairs([i,j,k,l,m],[])])>b:
32                       cuts.append([[i,j,k,l,m],[w,b]])
33        else:
34        break
35    s = newSolver('Sportsuschedule', True)
36    x = [[[s.IntVar(0,1,") if i<j else None
37           for _ in range(nbWeeks)]
38           for j in range(nbTeams)] for i in range(nbTeams-1)]
39    basic_model(s,Teams,nbTeams,nbWeeks,nbPerWeek,nbIntra,\
40              nbDiv,nbInter,cuts,x)
41    rc,Cal = s.Solve(),[]
42    if rc == 0:
43      Cal=[[(i,j) \
44            for i in range(nbTeams-1) for j in
              range(i+1,nbTeams)\
45            if SolVal(x[i][j][w])>0] for w in range(nbWeeks)]
46    return rc,ObjVal(s),Cal
```

The code starts with a loop on line 5 that will run a specific number of times, solving the model with fractional solutions. Line 11 is essentially all the constraints of Listing 7-17 (with the additional cuts) packaged in a procedure because we will need to use it multiple times, with fractional variables within the loop and finally with integer variables after the loop. After each solve, we consider tuples of teams, see if the sum of their decision variables exceeds the prescribed bound, and add their ordinal, along with the week under consideration and the bound to a list of cuts if it does.

Finally, we create an integer solver instance at line 35, add all the cuts previously found, and solve it for real. The routine to add the cuts is simply that shown in Listing 7-19.

Listing 7-19. Cuts Adding Routine

```
1  for t,w in cuts:
2      s.Add(s.Sum(x[p[0]][p[1]][w[0]] for p in pairs(t,[])) <= w[1])
```

In it, the `pairs` function generates all ordered pairs from ordered tuple t. Next, see Listing 7-20.

Listing 7-20. Ordered Pairs Generation

```
1  def pairs(tuple, accum=[]):
2      if len(tuple)==0:
3          return accum
4      else:
5          accum.extend((tuple[0],e) for e in tuple[1:])
6          return  pairs(tuple[1:],accum)
```

Let me stress that, contrary to TSP, where the subtour elimination constraints are required for the model to be valid, the constraints we are adding here are not required; they simply are added to nudge the solver in the right direction and accelerate the solution process. Therefore, for some solvers they will help tremendously. For others they will be useless; they might actually slow the whole process down. Without deep knowledge of the internal workings of a particular solver, the effect on runtime is nearly impossible to predict.[12] The point is that once a modeler is aware of this technique of relaxation tightening, he may easily try it on a particular combination of model-solver that is proving recalcitrant.

---

[12]Actually, even with deep internal knowledge, it may be near impossible to tell. Trying the approach to see if it works is orders of magnitude easier than reading the entrails of current integer solvers. Written in C, or worse C++, they have layers upon layers of complex cut generation routines with subtle interactions, not to mention cruft accumulated over years of development and debugging.

7.4.2 Variations

The variations on this model are multiple. Some of them affect the objective function (or, equivalently, are handled as soft constraints); some are hard constraints; some could be either.

- There could be a list of pairs (week, team, team) with the goal of having this specified match during that specified week.

- Instead of pushing intra-divisional matches towards the end, we could be asked to follow a specific pattern say Intra-Intra-Inter.

- We could be asked to spread out (or, bunch in) the multiple matches between pairs of teams.

- Instead of weeks, we could be asked to schedule on specific dates.

- We could add the concept of Home and Away games with the understanding that the number of Home games is fixed.

- There could also be a pattern of Home and Away games to follow. This might even be considered in the context of team cities with an eye towards a "reasonable" travel schedule. (What I am describing here is the addition of *multiple* TSP layers on top of an already difficult timetabling problem! Not for the faint of heart).

7.5 Puzzles

There is a long tradition in constraint programming of solving puzzles, mainly because it is amusing, though it is also educational. Solving puzzles using integer programming is not tried as often, yet it can be as amusing and educational if sometimes more difficult. Let's not be deterred by the difficulty. The tricks used in puzzles and the mental gymnastics used to model the problems can be used for "real" problems later.

7.5.1 Pseudo-Chess Problems

As a warm-up, let's consider a square chessboard of some specified size n on which we wish to place as many rooks as possible so that no rook is attacking any other.[13]

The question to answer is "Where to place the rooks to avoid attacks?" Therefore, an answer must be a set of positions occupied by rooks. Since the chessboard is square, an obvious formulation of the decision variable is a two-dimensional array of binary variables. So

$$x_{i,j} \quad \forall i \in \{0,\dots,n-1\}, \forall j \in \{0,\dots,n-1\},$$

Where if $x_{2,5}$ is one, then there is a rook at position 2, 5.

The objective function is simple: since we want to place as many rooks as possible, we sum our decision variables:

$$\sum_i \sum_j x_{i,j}$$

Now what would the constraints be to prevent a rook from attacking another? Rooks attack anything in the same column or row. We therefore need to have at most one rook per column and per row. This is a constraint we know well: the *one-out-of-n* constraint, implemented by the following:

$$\sum_i x_{i,j} \leq 1 \quad \forall j \in \{0,\dots,n-1\},$$
$$\sum_j x_{i,j} \leq 1 \quad \forall i \in \{0,\dots,n-1\}$$

---

[13]A rook attacks any piece in the same column or row, no matter the distance.

We have all we need. Let's translate this into executable code. The workhorse should be our k_out_of_n routine helped along by a couple of utility functions: one to extract all the row variables of a specified row and one to extract all the column variables. These utilities are seen in Listing 7-21.

Listing 7-21. Columns and Rows Extraction Utility (`puzzle.py`)

```
1  def get_row(x,i):
2     return [x[i][j] for j in range(len(x[0]))]
3  def get_column(x,i):
4     return [x[j][i] for j in range(len(x[0]))]
```

The main model creates a variable per board position and then forces at most one variable to be non-zero for each row and one for each column. The objective function sums all variables and the code returns a two-dimensional table of blanks and *R* to indicate where the rooks are positioned at optimality.

Listing 7-22. Maxrook model (`puzzle.py`)

```
1  def solve_maxrook(n):
2     s = newSolver('Maxrook',True)
3     x = [[s.IntVar(0,1,") for _ in range(n)] for _ in
       range(n)]
4     for i in range(n):
5       k_out_of_n(s,1,get_row(x,i),'<=')
6       k_out_of_n(s,1,get_column(x,i),'<=')
7     Count = s.Sum(x[i][j] for i in range(n) for j in
       range(n))
8     s.Maximize(Count)
9     rc = s.Solve()
10    y = [[['u','R'][int(SolVal(x[i][j]))]\
11         for j in range(n)] for i in range(n)]
12    return rc,y
```

Running Listing 7-22 on a board of size 8 can produce the solution shown in Table 7-16. It can solve boards of size 128 with the same ease.

Table 7-16. *An Optimal Solution to the Maxrook Puzzle*

	1	2	3	4	5	6	7	8
1		R						
2						R		
3								R
4							R	
5					R			
6			R					
7				R				
8	R							

So let's consider a slightly more difficult problem, the famous N-Queens. This is the same problem but we are asked to place queens instead of rooks. Queens attack on the diagonals as well as on the rows and columns. All we need is some convention for naming the diagonals and a function to extract them. To make this interesting, let's generalize our maxrook to a maxpiece accepting the type of chess piece to place. See Listing 7-23.

Listing 7-23. Diagonal Extraction Helper Functions (`puzzle.py`)

```
1  def get_se(x,i,j,n):
2      return [x[i+k % n][j+k % n] for k in range(n-i-j)]
3  def get_ne(x,i,j,n):
4      return [x[i-k % n][j+k % n] for k in range(i+1-j)]
```

We'll name diagonals either SE for southeast (to northwest) or NE for northeast (to southwest). Two utilities to extract the corresponding variables, get_se and get_ne, are shown in Listing 7-23.

The main model is shown in Listing 7-24. We can call this model for a board of set size equal to parameter n. The pieces can be queens, rooks, and bishops, indicated by the second parameter as Q, R, or B. A solution for queens and bishops is shown in Table 7-18. We also display in Table 7-17 the runtime of various instance size, normalized so that the time for $n = 2$ is one. These values are to be taken with a large dose of salt, as they are solver dependent. They nevertheless suggest that the models presented do not suffer from the exponential explosion that a purely combinatorial solver might encounter.

Listing 7-24. Maxpiece General Model (`puzzle.py`)

```
1  def solve_maxpiece(n,p):
2    s = newSolver('Maxpiece',True)
3    x = [[s.IntVar(0,1,") for _ in range(n)] for _ in range(n)]
4    for i in range(n):
5      if p in ['R' ,'Q']:
6        k_out_of_n(s,1,get_row(x,i),'<=')
7        k_out_of_n(s,1,get_column(x,i),'<=')
8      if p in ['B', 'Q']:
9        for j in range(n):
10          if i in [0,n-1] or j in [0,n-1]:
11            k_out_of_n(s,1,get_ne(x,i,j,n),'<=')
12            k_out_of_n(s,1,get_se(x,i,j,n),'<=')
13    Count = s.Sum(x[i][j] for i in range(n) for j in range(n))
14    s.Maximize(Count)
15    rc  = s.Solve()
16    y=[[['u',p]\
17      [int(SolVal(x[i][j]))] for j in range(n)] for i in range(n)]
18    return rc,y
```

Table 7-17. *Runtime for Increasing Board Size*

8	1
16	3
32	9
64	43
128	169
256	870
512	6318

Table 7-18. *An Optimal Solution to the N-Queens and Max Bishops Puzzles*

	1	2	3	4	5	6	7	8		1	2	3	4	5	6	7	8
1				Q					1			B	B			B	
2								Q	2	B							
3	Q								3								B
4					Q				4								B
5							Q		5	B							
6		Q							6	B							
7						Q			7								B
8			Q						8	B	B				B	B	B

Note that there is an obvious generalization to any piece, since each occupied position on the board describes a set of positions, hence variables and the sum of them must be one.

Looking at the example solutions, two obvious questions arise: Can we get all solutions? Can we get "interesting" solutions? We will leave the first question aside for now and consider the second. What would constitute an interesting solution? Maybe one that exhibits some symmetry. We could try to minimize the sum of the distances between pieces, or the maximum distance, and see what happens. The reader is invited to implement these modifications. We now take our leave of pseudo-chess.

7.5.2 Sudoku

The Sudoku puzzles is as follows: given a 9 by 9 grid, partially filled with numbers in the range 1 to 9, fill the rest of the grid such that

- Every row contains all numbers 1 to 9.

- Every column contains all numbers 1 to 9.

- Every 3 by 3 disjoint subgrid contains all numbers 1 to 9.

We can represent a solution by specifying, for every grid position, which number is in there. So a simple decision variable is

$$x_{i,j} \in \{1,\ldots,9\} \quad \forall i \in [1,2,3] \forall j \in [1,2,3]$$

The constraints are interesting. Each one of them is of the form "Given a given set of nine positions, all numbers from 1 to 9 must appear." In constraint programming, this requirement is handled by a single functional call, usually named all_different. We will create a simplified equivalent constraint for Sudoku and improve it for the next puzzle.

For every one of our variable $x_{i,j}$ we will create an array v_k^{ij} of length 9 of binary variables. Each of these is an indicator for the value k. So that we add the constraint

$$x_{i,j} = v_1^{ij} + 2v_2^{ij} + 3v_3^{ij} + \ldots + 8v_8^{ij} + 9v_9^{ij} \tag{7.14}$$

Then, for every set *S* of variables that need to be all different, we will ensure that the sum of the corresponding indicator variables sums to one.

This is a feasibility problem so no objective function is required. Let's create an executable model. We leverage our previously defined get_row and get_column to which we are adding a get_subgrid. See Listing 7-25.

Listing 7-25. Some Helper Functions for Sudoku (puzzle.py)

```
1  def get_subgrid(x,i,j):
2     return [x[k][l] for k in range(i*3,i*3+3)\
3                     for l in range(j*3,j*3+3)]
4  def all_diff(s,x):
5     for k in range(1,len(x[0])):
6        s.Add(sum([e[k] for e in x]) <= 1)
```

The model implemented in Listing 7-26 accepts a grid of data with either a number or None to indicate that it must be filled. Most of the work is to create a clean set of decision variables in the loop from line 3 to line 14: a three-dimensional array indexed by the position on the grid for the first two dimensions. At index 0 in the third dimension is our real decision variable, holding the value that the grid will hold (either because it is data or after the solution process); each other index from 1 to 9 holds the corresponding indicator variable. As we create the variables we add the value constraint of equation (7.14) at line 9.

After the variable declaration we call the all_diff function of each row, column, and subgrid. This function is a simple k_out_of_n for each value in the range 1 to 9.

Listing 7-26. Sudoku Model (puzzle.py)

```
1   def solve_sudoku(G):
2     s,n,x  =  newSolver('Sudoku',True),len(G),[]
3     for i in range(n):
4       row=[]
5       for j in range(n):
6         if G[i][j] == None:
7          v=[s.IntVar(1,n+1,")]+[s.IntVar(0,1,")\
8                           for _ in range(n)]
9          s.Add(v[0] == sum(k*v[k] for k in range(1,n+1)))
10         else:
11          v=[G[i][j]]+[0 if k!=G[i][j] else 1\
12                    for k in range(1,n+1)]
13         row.append(v)
14       x.append(row)
15     for i in range(n):
16       all_diff(s,get_row(x,i))
17       all_diff(s,get_column(x,i))
18     for i in range(3):
19       for j in range(3):
20         all_diff(s,get_subgrid(x,i,j))
21     rc  = s.Solve()
22     return rc,[[SolVal(x[i][j][0]) for j in range(n)]\
23               for i in range(n)]
```

Finally, we return the grid values, not the 800 or so indicator variables. As an example, see Table 7-19. The data are in bold.[14]

Table 7-19. *Solution to a Sudoku Puzzle*

1	2	5	8	**3**	7	**6**	9	4
4	**7**	6	2	1	9	8	3	**5**
9	3	8	4	**6**	5	7	2	1
8	6	3	**7**	4	**1**	9	5	2
2	5	1	6	9	3	**4**	7	8
7	4	9	**5**	8	2	1	6	3
5	8	4	**9**	2	6	3	**1**	**7**
6	1	**2**	3	7	4	5	**8**	9
3	9	7	1	5	8	2	4	6

7.5.3 Send More Money!

Here is a crypt-arithmetic puzzle, famous in the constraint programming community: Replace each of the letters S, E, N, D, M, O, R, Y, with a distinct digit from 0 to 9 such that the following sum is correct:

```
SEND  +  MORE  =  MONEY
```

There are two high-level constraints in this puzzle. The first is arithmetic: we need the equation to hold. We can do this by decomposing each integer into its place-value. SEND is a four-digit number (presumably in base ten) so it really is

$$S*1000 + E*100 + N*10 + D*1$$

---

[14]I have run this code on over 20,000 puzzles. In most cases, the model runs in a small fraction of a second; occasionally it will take a few seconds.

Each of MORE and MONEY is handle the same way. Then we constrain the equation to hold.

The second constraint is the requirement that all letters receive a distinct digit. Here is another case where we could profitably use an all-different constraint, so let's generalize what we did for the pseudo-chess models so that we can invoke all-different in any model. Our trick relies on each variable having an array of associated indicator variables, one for each potential integer value. So, in addition to our previously defined constraint, which we will slightly generalize, we need a routine for variable creation. This is the intent of newIntVar of Listing 7-27.

Listing 7-27. The General All-Different Structure and Constraint (puzzle.py)

```
1  def newIntVar(s, lb, ub):
2    l=ub-lb+1
3    x=[s.IntVar(lb, ub, ")]+[s.IntVar(0,1,") for _ in range(l)]
4    s.Add(1 == sum( x[k] for k in range(1,l+1)))
5    s.Add(x[0] == sum((lb+k-1)*x[k] for k in range(1,l+1)))
6    return x
7  def all_different(s,x):
8    lb=min(int(e[0].Lb()) for e in x)
9    ub=max(int(e[0].Ub()) for e in x)
10   for v in range(lb,ub+1):
11     all = []
12     for e in x:
13       if e[0].Lb() <= v <= e[0].Ub():
14         all.append(e[1 + v - int(e[0].Lb())])
15     s.Add(sum(all)  <=  1)
16 def neq(s,x,value):
17   s.Add(x[1+value-int(x[0].Lb())] == 0)
```

We notice that, although unstated, there is an additional assumption on S and M: they cannot take on value 0 if the numbers are truly four and five digits long. So we should constrain them to be non-zero. Not coincidentally, the data structure we have chosen for the all_different implementation allows us to trivially create a disequality as you see in function neq of Listing 7-27.

Armed with this, we can now solve the puzzle. The implementation is shown in Listing 7-28. Its solution is shown in Table 7-20.

Listing 7-28. Send More Money (`puzzle.py`)

```
1  def solve_smm():
2    s = newSolver('Sendumoreumoney',True)
3    ALL = [S,E,N,D,M,O,R,Y] = [newIntVar(s,0,9) for
       k in range(8)]
4    s.Add( 1000*S[0]+100*E[0]+10*N[0]+D[0]
5          + 1000*M[0]+100*O[0]+10*R[0]+E[0]
6          == 10000*M[0]+1000*O[0]+100*N[0]+10*E[0]+Y[0])
7    all_different(s,ALL)
8    neq(s,S,0)
9    neq(s,M,0)
10   rc = s.Solve()
11   return rc,SolVal([a[0] for a in ALL])
```

The reader can verify that the equation holds ($9567 + 1085 = 10652$).

Table 7-20. *Solution to the Send More Money Puzzle*

S	E	N	D	M	O	R	Y
9	5	6	7	1	0	8	2

7.5.4 Ladies and Tigers

Raymond Smullyan in "The Lady or the Tiger"[15] presents a number of logic puzzles. One chapter culminates in the following:

A prisoner is faced with nine doors, one of which he must open. Behind one door awaits a lady; behind the others, a tiger, if anything. One assumes that the prisoner prefers opening the lady's door to an empty room, which is preferable to a tiger's den. What turns this into a logic puzzle is that on each door is posted a logic statement (it can therefore be either true or false). Statements on rooms with tigers are false. The statement on the Lady's room is true.

- Door 1: The lady is in an odd-numbered room.

- Door 2: This room is empty.

- Door 3: Either sign 5 is right or sign 7 is wrong.

- Door 4: Sign 1 is wrong.

- Door 5: Either sign 2 or sign 4 is right.

- Door 6: Sign 3 is wrong.

- Door 7: The lady is not in room 1.

- Door 8: This room contains a tiger and room 9 is empty.

- Door 9: This room contains a tiger and sign 6 is wrong.

---

[15]Raymond M. Smullyan, *The Lady or the Tiger, and Other Logic Puzzles* (Mineola, New York: Dover Publications, 2009).

Where is the lady?

To know where the lady awaits, we may need to know where the tigers are. So a reasonable decision variable, given a set $R = \{1, \dots, 9\}$ of rooms and a set $B = \{1, 2, 3\}$ of beasts (say 1 for empty, 2 for lady, and 3 for tiger) would be

$$r_i \in B \quad \forall i \in R$$

So that a lady in room 5 would be indicated by $r_5 = 2$ and a tiger in room 4 would be $r_4 = 3$. A statement about the lady being in an odd numbered room is easy to accommodate if we declare our variables using our newIntVar function. The associated array of indicator variables is the perfect tool. For the sake of presentation, let's assume that for each r_i variable we have an array $r_{i,j}$ of indicator variables for $j \in B$.

Now for the logic part. We are given statements that can be true or false and their truth value influences the constraints. If we introduce a binary variable for each statement, then we can use our reify logic to associate the variable to each constraint. So let's introduce

$$s_i \in \{0,1\} \quad \forall i \in R,$$

so that $s_2 = 1$ will mean that the statement on door 2 is true.

The executable model is shown in Listing 7-29. We will decompose it one constraint at a time.

Listing 7-29. Lady or Tiger Model (puzzle.py)

```
1  def solve_lady_or_tiger():
2      s = newSolver('Ladyuorutiger', True)
3      Rooms = range(1,10)
4      R = [None]+[newIntVar(s,0,2) for _ in Rooms]
5      S = [None]+[s.IntVar(0,1,") for _ in Rooms]
6      i_empty,i_lady,i_tiger = 1,2,3
```

```
7     k_out_of_n(s,1,[R[i][i_lady] for i in Rooms])
8     for i in Rooms:
9       reify_force(s,[1],[R[i][i_tiger]],0,S[i],'<=')
10      reify_raise(s,[1],[R[i][i_lady]],1,S[i],'>=')
11    v=[1]*5
12    reify(s,v,[R[i][i_lady] for i in range(1,10,2)],1,S[1],'>=')
13    reify(s,[1],[R[2][i_empty]],1,S[2],'>=')
14    reify(s,[1,-1],[S[5],S[7]],0,S[3],'>=')
15    reify(s,[1],[S[1]],0,S[4],'<=')
16    reify(s,[1,1],[S[2],S[4]],1,S[5],'>=')
17    reify(s,[1],[S[3]],0,S[6],'<=')
18    reify(s,[1],[R[1][i_lady]],0,S[7],'<=')
19    reify(s,[1,1],[R[8][i_tiger],R[9][i_empty]],2,S[8],'>=')
20    reify(s,[1,-1],[R[9][i_tiger],S[6]],1,S[9],'>=')
21    rc  = s.Solve()
22    return rc,[SolVal(S[i]) for i in Rooms],\
23      [SolVal(R[i]) for i in Rooms]
```

At line 3 we define the range of integers that identify each door. Since the problem is stated with rooms numbered from one, we will comply instead of renumbering from zero as we usually do. In order to index from one, we create the decision variables on the next two lines as arrays where the first element contains None. We then define, at line 6, some constants to access the indicator variables of each room.

On line 7 we ensure that there is exactly one lady.

All the other constraints involve a relation between a statement variable S and a logical statement, hence our reify functions will prove invaluable. The first one is that if a room contains a tiger, its statement is false. The statement "If room i contains a tiger then statement i is false." is in the wrong direction for us. We could write a new function but is simpler to use the contra-positive and state "If statement i is true, there is no tiger behind door i." This is an instance of a Boolean true enforcing a constraint, which line 9 implements.

259

The next constraint is "A room with a lady has a true statement on its door." and it is in the right direction for us: a satisfied constraint raising a Boolean, as implemented at line 10.

"The lady is in an odd-numbered room." is simple. We need to sum the i_lady indicator variables on odd doors and set them above one if and only if statement one is true. The indices of odd rooms is obtained by range(1,10,2) and the inequality

```
R[1][i_lady]+R[3][i_lady]+R[5][i_lady]+R[7][i_lady]+R[9][i_
lady]  >=  1
```

is reified to S[1] at line 12.

"This room is empty." is a simple reification to S[2] of

```
R[2][i_empty]  >=  1
```

at line 13. The reader might wonder why not use equality instead of inequality. The reason is that we know, having written the constraint reify, that equalities are more complex, introducing more auxiliary constraints and or variables. If we are certain, as we are here, that an inequality is sufficient, it is preferable to use one.

"Either sign 5 is right or sign 7 is wrong." is a disjunction of binary variables with the minor difficulty that the second is negated. The transformation to an algebraic statement is mechanical if we know how to negate Booleans and implement disjunctions. We have seen the disjunctions often. Something like $x_1 \lor \ldots x_n$ gets implemented as $\sum_i x_i \geq 1$. The negation of x_i is handled by replacing x_i by $1 - x_i$. Therefore, in our case we need

```
S[5] + (1-S[7]) >= 1
```

which simplifies to

```
S[5] - S[7] >= 0
```

reified to S[3] at line 14.

"Sign 1 is wrong." is S[1]==0 reified to S[4] at line 15. "Sign 3 is wrong." is handled identically at line 17.

"Either sign 2 or sign 4 is right." is a simple disjunction so the usual transformation to addition applies and is reified to S[5] at line 16.

"The lady is not in room 1." needs to reify to S[7] the constraint

```
R[1][i_lady]  <=  0
```

as at line 18.

"This room contains a tiger and room 9 is empty." is interesting. The substatements are

```
R[8][i_tiger]  >=  1
```

and

```
R[9][i_empty]  >=  1
```

The conjunction is handled by summing the right sides and the left sides to create

```
R[8][i_tiger]  +  R[9][i_empty]  >=  2
```

reified to S[8] at line 19.

Finally, "This room contains a tiger and sign 6 is wrong." contains

```
R[9][i_tiger]  >=  1
```

and

```
S[6] <= 0
```

We transform the latter into

```
- S[6] >= 0
```

And then sum the two form the conjunction, reified to S[9] at line 20.

And we are done in the sense that we can find a solution, for instance the first solution of Table 7-21. But are there additional solutions and if so, how to find them? In this case,[16] it is simple since our only real goal is to find the lady. Our first solution has the lady in room 1, so we can simply add a constraint preventing the lady from being in room 1, for instance

```
s.Add(R[1][i_lady]  ==  0)
```

We will get another solution of one exists or else the solver will indicate that the problem is infeasible. We can proceed thus until we exhaust all solutions if need be. The second solution of Table 7-21 is one such additional solution. (Interestingly, this second solution is the unique solution if we add a constraint stating "Room 8 is not empty.")

Table 7-21. *Two Solutions to the Lady or Tiger Puzzle*

1	The lady is in an odd-numbered room.	T	Lady	T	
2	This room is empty.	T		F	Tiger
3	Either sign 5 is right or sign 7 is wrong.	T		F	
4	Sign 1 is wrong.	F		F	
5	Either sign 2 or sign 4 is right.	T		F	
6	Sign 3 is wrong.	F		T	
7	The lady is not in room 1.	F		T	Lady
8	This room contains a tiger and room 9 is empty.	F		F	Tiger
9	This room contains a tiger and sign 6 is wrong.	F		F	Tiger

---

[16]In general, for integer programs, it is very difficult to find all solutions, but it is possible for many practical cases.

7.6 Quick Reference for OR-Tools MPSolver in Python

This is by no means a complete reference, but it is enough for all the models described in the book. Also described are the wrapper functions used to simplify model writing.

To use the LP and IP solvers, a model must start with the code in Listing 7-30.

Listing 7-30. Library Declaration

```
from linear_solver import pywraplp
```

A solver instance is created by the code in Listing 7-31.

Listing 7-31. Creation of a Solver Instance in OR-Tools

```
s = pywraplp.Solver(NAME,pywraplp.Solver.TYPE)
```

Here s is the returned solver, NAME is any string, and TYPE is one of

GLOP_LINEAR_PROGRAMMING	LP
CLP_LINEAR_PROGRAMMING	LP
GLPK_LINEAR_PROGRAMMING	LP
SULUM_LINEAR_PROGRAMMING	LP
GUROBI_LINEAR_PROGRAMMING	LP
CPLEX_LINEAR_PROGRAMMING	LP
SCIP_MIXED_INTEGER_PROGRAMMING	MIP
GLPK_MIXED_INTEGER_PROGRAMMING	MIP
CBC_MIXED_INTEGER_PROGRAMMING	MIP

SULUM_MIXED_INTEGER_PROGRAMMING	MIP
GUROBI_MIXED_INTEGER_PROGRAMMING	MIP
CPLEX_MIXED_INTEGER_PROGRAMMING	MIP
BOP_INTEGER_PROGRAMMING	IP (Binary)

All the models in the book use GLOP for LP problems and CBC for all MIPs. Listing 7-32 shows how the current wrapper is set.

Listing 7-32. Creation of a Solver Instance via Wrapper

```
s = newSolver(NAME,[False|True])
```

Note that False is the default and returns an LP solver instance (GLOP) and True returns a MIP solver (CBC).

To a solver instance we add decision variables by the code in Listing 7-33 for continuous variables or by the code in Listing 7-34 for integer variables where VAR is the returned variable object and NAME is any string. Names must be unique within a solver instance. Given the empty string, an automatic unique name will be internally generated, which is a precious feature, especially for routines that will be repeatedly reused within a solver instance. The range is described by LOW, any number, or -solver.infinity() and by HIGH, any number larger than LOW, or solver.infinity(). It is a good rule of thumb to restrict the range as much as possible.

Listing 7-33. Continuous Decision Variable Declaration via OR-Tools

```
var = s.NumVar(LOW,HIGH,NAME)
```

Listing 7-34. Integer Decision Variable Declaration via OR-Tools

```
VAR = s.IntVar(LOW,HIGH,NAME)
```

The easiest way to create an array of decision variables is shown in Listing 7-35.

Listing 7-35. Decision Variable Array Declaration Example

```
x = [s.NumVar(LOW,HIGH,") for _ in range(N)]
```

Note that N is the number of elements required in the array. Arrays are, of course, indexed starting at zero. This is similar for high-dimensional arrays. For example, an M by N matrix is created by the code in Listing 7-36.

Listing 7-36. Decision Variable 2-D Array Declaration Example

```
m =[[s.NumVar(LOW,HIGH,") for _ in range(N)] for _ in range(M)]
```

After variable declarations, constraints follow. The simplest constraint declaration is shown in Listing 7-37.

Listing 7-37. Generic Constraint Declaration via OR-Tools

```
s.Add(REL)
```

Note that REL is (almost) any linear algebraic relation using decision variables, numbers, the arithmetic operators +,-,*,/ and the equality and inequality relationals. For an example, see Listing 7-38.

Listing 7-38. Simple Algebraic Constraints in OR-Tools

```
s.Add(2 * x[12] + 30 * x[13] <= 100)
s.Add(25 == x[100] - x[101])
s.Add(x[99] >= x[100])
```

Never use the strict inequalities. For continuous variables, they make no sense; for integer variables, they can trivially be changed to the non-strict inequalities by adding one. Also remember that we can never use products of decision variables.

A useful helper function is the sum, invoked as shown in Listing 7-39.

Listing 7-39. Sum Operator in OR-Tools

```
s.Sum(LIST)
```

Here LIST is any list (or tuple) of decision variables. This can be used, for instance, as shown in Listing 7-40.

Listing 7-40. Examples of Sum in OR-Tools

```
s.Add(s.Sum(x)   <=   100)
Add(s.Sum(m[i][j] for i in range(M) for j in range(N))<= 100)
```

In addition to the constraints, a model usually has an objective function of either of the forms shown in Listing 7-41.

Listing 7-41. Objective Function Declarations in OR-Tools

```
s.Maximize(EXPR)
s.Minimize(EXPR)
```

Here EXPR is any linear algebraic expression in the decision variables. To invoke the solver on the model created, see Listing 7-42.

Listing 7-42. Solver Invocation in OR-Tools

```
rc = s.Solve()
```

Here `rc` is the returned value and is zero if all goes well. It can be one of

```
OPTIMAL
FEASIBLE
INFEASIBLE
UNBOUNDED
ABNORMAL
NOT_SOLVED
```

So, to be pedantic, one should check the return value against those defined constants but the programmers at Google have followed the decades-old tradition of returning zero when all goes well.[17]

After a solve, one accesses the optimal value and optimal solution via the code in Listing 7-43.

Listing 7-43. Optimal Value and Optimal Solution

```
value = s.Objective().Value()
varval = var.SolutionValue()
```

They are wrapped in the helper functions shown in Listing 7-44.

Listing 7-44. Wrapper Functions for Optimal Value and Solutions

```
value = ObjVal(s)
xval = SolVal(x)
```

The returned variable `xval` will have the same dimensions as the parameter `x`.

In addition, the wrapper library provides the following higher-level constraints shown in Listing 7-45.

---

[17] I believe the tradition started with Dennis Ritchie and Unix as simplifying medication against the Multics headache-inducing complexities.

Listing 7-45. High-Level Constraints

```
l = k_out_of_n(s,k,x,rel='==')
l = sosn(solver,k,x,rel='<=')
delta = reify_force(s,a,x,b,delta=None,rel='<=',bnds=None)
delta = reify_raise(s,a,x,b,delta=None,rel='<=',bnds=None,eps=1)
delta = reify(s,a,x,b,d=None,rel='<=',bnds=None,eps=1)
```

Here,

- k_out_of_n adds the necessary constraints to solver s so that exactly, at least, or at most (depending on rel) k (a positive integer) variables from the list x are allowed to be non-zero. It returns l, an array of binary variables of the same length as x.

- sosn adds the necessary constraints to solver s so that exactly, at least, or at most (depending on rel) k adjacent variables from the list x are allowed to be non-zero. It returns l, an array of binary variables of the same length as x.

- reify_force adds the necessary constraints to solver s so that the relation $\sum_i a_i x_i \approx b$ is forced (for relation $\approx$ determined by rel) when d (a binary integer variable) is one. This last variable need not be declared ahead of the call to reify_force. It is returned whether created internally or not.

- reify_raise implements the opposite implication to force.

- reify calls both force and raise to implement an "if and only if" condition.

It also provides the helper function.

Next is Listing 7-46.

Listing 7-46. Bounds Extraction

```
bounds_on_box(a,x,b)
```

Here bounds_on_box finds the smallest and largest possible value of $\sum a_i x_i - b$ on the domain of variable x.

A final constraint is the all_different, shown in Listing 7-47.

Listing 7-47. All_different Predicate

```
all_different(s,x)
```

Here s is the solver and x is a collection of decision variables of integer type. They all should have lower and upper bounds. It will enforce that no two of the decision variables will have the same value.

Index

A

Abstract approach, 79
Amphibian coexistence
  aquarium, 7
  executes, 16
  model, 13–15
  modeling languages, 11–12
  MPSolver, 12
  OR-Tools library, 12
  preys, 7
  run a model, 17–18
  solution, 17
  three-step approach, 7–10
Attendant theory, 4

B

Bin packing
  constraints, 139–140
  decision variables, 139
  description, 137
  executable model
    array of weights, 142
    bound-trucks, 142
    constraints, 143
    decision variables, 143, 145
    degeneracy, 144

    optimal package assignments, 143
    optimal package assignments, symmetry-breaking, 147
    optimal truck loads, symmetry-breaking, 147
    over-constraining, 146
    packages and trucks, 145–146
    packages list, weights, 141
    simple heuristic to bound, trucks, 148
    symmetry-breaking constraints, 144, 146–147
    symmetry_break parameter, 143–145
    tab, 143
    truck numbers, 143
    truck selection variables, 144
  objective, 141
  packages, 138
  trucks, weight capacity, 138
  variations
    capital budgeting, 149
    knapsack, 149
    objective function, 149
    symmetry-breaking, 148

T, U, V, W, X, Y, Z

Get the eBook for only $5!

Why limit yourself?

With most of our titles available in both PDF and ePUB format, you can access your content wherever and however you wish—on your PC, phone, tablet, or reader.

Since you've purchased this print book, we are happy to offer you the eBook for just $5.

To learn more, go to http://www.apress.com/companion or contact support@apress.com.

Apress®

Printed in the United States
By Bookmasters